NDT in Canada 2018

Halifax, Nova Scotia, Canada
19 - 21 June 2018

ISBN: 978-1-5108-7187-8

Printed from e-media with permission by:

Curran Associates, Inc.
57 Morehouse Lane
Red Hook, NY 12571

Some format issues inherent in the e-media version may also appear in this print version.

Printed by Curran Associates, Inc. (2018)

For permission requests, please contact NDT.net
at the address below.

NDT.net
Rheintalstr 3
53498 Bad Breisig
Germany

Phone: + 49 2633 4756670
Fax: + 49 3212-1030513

info@ndt.net

Additional copies of this publication are available from:

Curran Associates, Inc.
57 Morehouse Lane
Red Hook, NY 12571 USA
Phone: 845-758-0400
Fax: 845-758-2633
Email: curran@proceedings.com
Web: www.proceedings.com

TABLE OF CONTENTS

Preliminary Investigation of Non-Destructive Evaluation Techniques for Parts Produced by Additive Manufacturing

Jeffrey McIsaac, General Manager AMIC
Mohammadarmaan Bandi, AMIC
Doug Whitely, CINDE

CINDE CANADIAN INSTITUTE FOR NON-DESTRUCTIVE EVALUATION

MOHAWK IDEAWORKS ADDITIVE MANUFACTURING INNOVATION CENTRE

NDT in Canada
NDTC 2018
Canada's NDT Conference
June 19 - 21
Halifax Convention Centre
Halifax, Nova Scotia

What is Additive Manufacturing (AM) ?

- Additive manufacturing is a technology which build up a component by using digital 3D design data, usually joining layer-upon-layer of deposited material.
- Also know as "3D printing"

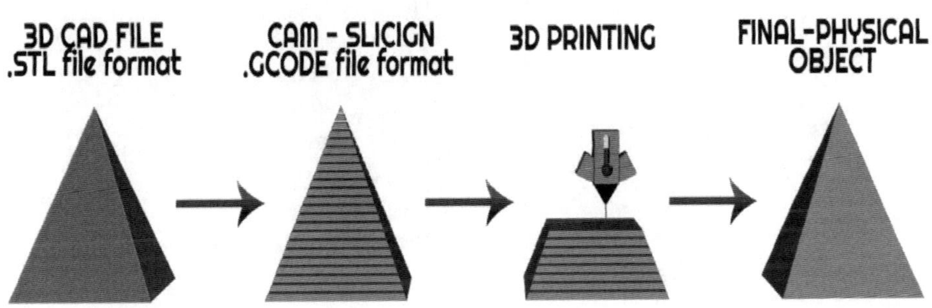

Figure 1: General principle of additive manufacturing

Source: http://my3dconcepts.com/explore/how-3d-printing-works/

Advantages of Additive Manufacturing

- Lightweight parts
- Complex geometries
- Mass customization
- Rapid prototyping
- Functional integration
- Design flexibility

Powder Bed Fusion(PBF) Additive Manufacturing

- PBF technology produces a solid part using a thermal source, such as laser or electron-beam, to sinter and join metal or plastic powder particles together, one layer at a time
- Metal Manufacturing- Stainless Steel, Aluminium, Titanium and other alloys

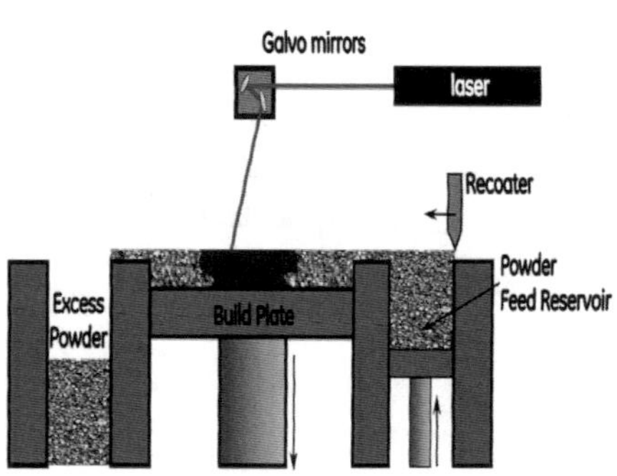

Figure 2: The Powder Bed fusion process

Figure 3: EOS M280 Direct Metal Laser Sintering system at Mohawk College's Additive Manufacturing Innovation Centre

Impact of AM on Non-Destructive Evaluation(NDE)

- AM's unique part properties and part microstructures with complexity in design raises challenges for NDE.
- Key questions-
  - *How would NDT factor into such a new manufacturing process?*
  - *Would there be any differences between traditionally manufactured components and additive manufacturing components that would influence inspection?*
  - *What standards would guide NDT technicians?*

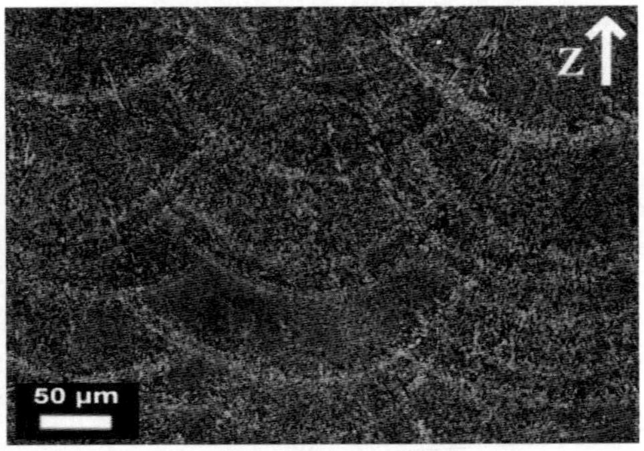

Figure 4: Optical micrographs of AlSi10Mg samples produced by means of DMLS

Figure 5: Mold block designed with internal, conformal cooling channels

Test Coupons

- Initial test blocks were produced in EOS M280 System from MS1- maraging tool steel and Ti64- Titanium alloy with same geometry
- Some challenges were experienced while manufacturing blocks from Ti64, deformation of the components due to high residual stresses

Figure 4: Coupons produced using MS1

Figure 5: Coupons produced using Titanium alloy showing delamination at the substrate interface due to residual stress

Test Coupons

- More advanced geometry created with internal defects.
- 5-step calibration block with various flaws produced in AlSi10Mg- Aluminum alloy.

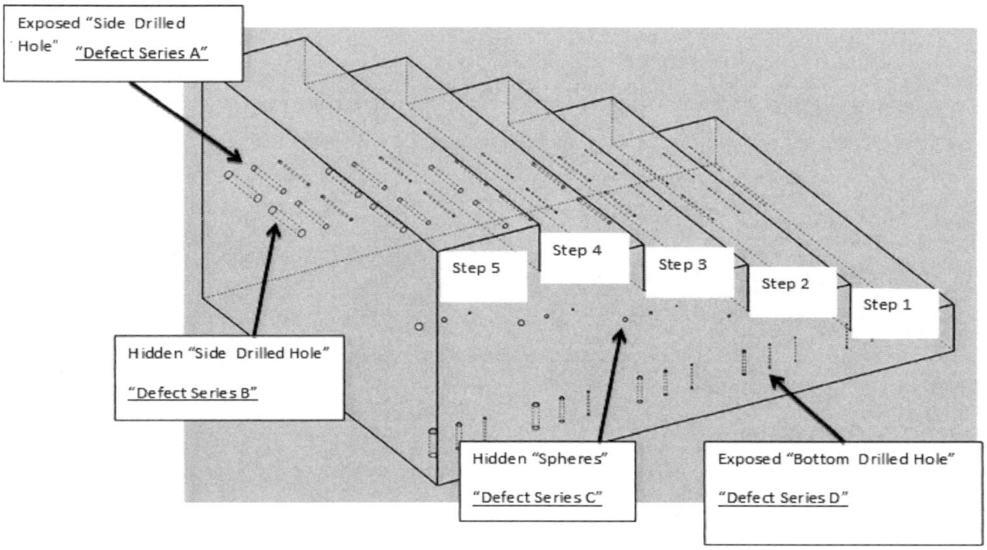

Figure 6: Step calibration block produced using AlSi10Mg

How did CINDE become involved?

- Mohawk College's Additive Manufacturing Group and specifically, Jeffrey McIsaac came to CINDE seeking some answers when something went wrong during the manufacture of a part.
- Specific questions were:
  - Can you help us determine if there are any additional internal "defects" that are not visible or on the surface.
  - How deep does this "defect" go?
- Unfortunately due to the parts geometry and thickness we did not have the equipment to help the AM group.
- It did however open the discussion as to how could we assist the AM Group.
- This led to the following questions for CINDE.

"Original Part" that started it all

From an NDE perspective

- Limited knowledge of the process of Additive Manufacturing. I called it 3D printing.
- Very limited knowledge of the type/nature or kind of discontinuity to be detected.
- Where to "look" for these discontinuities?
- How would the discontinuities present themselves?
- What would be the best "method" for detection?
- This lack of knowledge is what led to the first sample blocks being simple rectangular shapes of limited thickness.

"Needle in a Haystack"

CINDE
CANADIAN INSTITUTE FOR NON-DESTRUCTIVE EVALUATION

NDT in Canada 2018 | June 19-21 | Halifax, NS

9

Test Coupons

- Initial test blocks were produced in EOS M280 System from MS1-maraging tool steel and Ti64-Titanium alloy with same geometry
- These blocks were "cleaned up" prior to the inspection and the delamination was removed from the Titanium alloy blocks.

Figure 9: Coupons produced using MS1

Figure 10: Delamination at the substrate interface due to residual stress was removed prior to NDE inspection. Also note the surface roughness which may lead to difficult surface inspection for nonferrous materials. This resulted in the test blocks not being the intended "height".

CINDE
CANADIAN INSTITUTE FOR NON-DESTRUCTIVE EVALUATION

NDT in Canada 2018 | June 19-21 | Halifax, NS

10

Radiographic Set-up for MS1- maraging tool steel and Ti64-Titanium alloy

- Radiation source was aligned top center.
- FFD/SFD were 914 mm for x-ray and 500 mm for IR^{192}.
- Film Types were Type II for MS1-maraging tool steel and Type I for Ti64-Titanium alloy.
- IQI's (both Wire and Plaque type) were source side.
- Ultrasonic inspection was contact with inspection from 2 surfaces with 90 deg rotation.

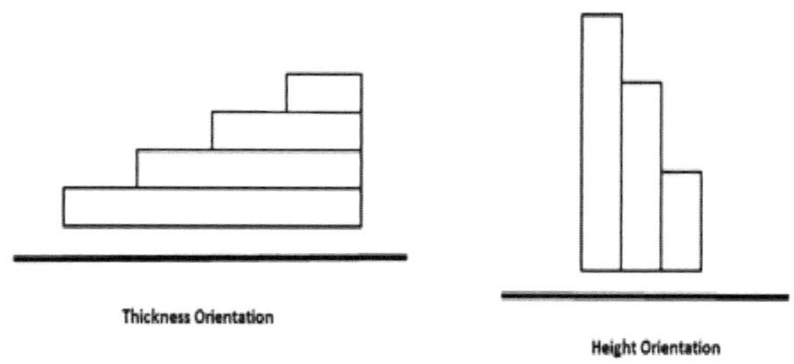

Figure 11: Radiographic set-up

CINDE
CANADIAN INSTITUTE FOR NON-DESTRUCTIVE EVALUATION

Chart 1 - Radiographic Results for MS1- maraging tool steel

- With the exception of Exposures 7 and 14 Radiographic sensitivity was excellent with contrast and detail achieving a 2- 1T with the Plague IQI with similar or greater sensitivity measured on the Wire IQI.
- Exposures 7 and 14 achieved only a 2- 4T sensitivity with the Plaque IQI. The cause for this result?
- Ultrasonic and Penetrant inspection revealed no discontinuities.

Chart 1 - EOS Maraging Steel MS1

Exposure	Material Thickness (in.)	Directtion	kVp/Source	SFD	mAs or Cimin	Plaque IQI # Designation, Viewed		ASTM IQI Designation, Wire viewed	
7	0.5	Thickness	200 kVp	914 mm	250	#10	4T	B	#6
8	1	Thickness	260 kVp	914 mm	500	#20	1T	B	#6
9	1.5	Thickness	260 kVp	914 mm	960	#30	2T	B	#8
10	2	Thickness	280 kVp	914mm	1100	#40	2T	B	#8
11	1	Height	260 kVp	914mm	400	#20	2T	B	#6
12	1.5	Height	260 kVp	914mm	1200	#30	1T	B	#8
13	2	Height	280kVp	914mm	1200	#40	2T	B	#9
14	0.5	Thickness	IR^{191}	20"	82	#10	4T	A	#6
15	1	Thickness	IR^{192}	20"	140	#20	1T	B	#8
16	1.5	Thickness	IR^{193}	20"	290	#30	2T	B	#8
17	2	Thickness	IR^{192}	20"	576	#40	1T	B	#8

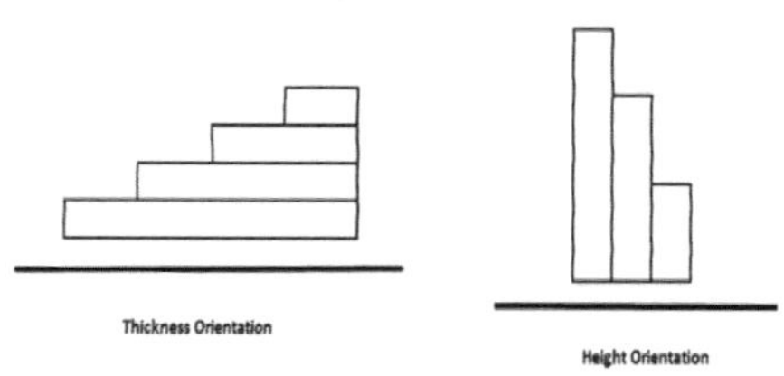

CINDE
CANADIAN INSTITUTE FOR NON-DESTRUCTIVE EVALUATION

Chart 2 - Radiographic Results for Ti64-Titanium alloy

- With the exception of Exposure 9 radiographic sensitivity was excellent with contrast and detail achieving a 2- 1T/2T with the Plague IQI with similar or greater sensitivity measured on the Wire IQI.
- Exposures 9 achieved only a 2- 4T sensitivity with the Plaque IQI and with IR 192 as the radiation source. Also note the "A" Wire IQI was not resolved in this exposure as well.
- Ultrasonic and Penetrant inspection revealed no discontinuities.

Chart 2 - Titanium Alloy

Exposure	Material Thickness (in.)	Directtion	kVp/Source	SFD	mAs or Cimin	Plaque IQI # Designation, Viewed		ASTM IQI Designation, Wire viewed	
All film is Type 1 with .005FR and .010 B									
1	0.5	Thickness	160	914 mm	600 mAs	10	1T	A	#3
								B	#6
2	1.1875	Thickness	180	914 mm	600	20	1T	A	Nil
								B	#6
3	1.5	Thickness	200	914 mm	900	30	1T	B	#6
4	2	Thickness	220	915 mm	900	40	1T	B	#6
5	0.5	Thickness	IR^{192}	20"	196	10	2T	A	Nil
								B	#6
6	1.1875	Thickness	IR^{192}	20"	245	20	2T	A	Nil
								B	#7
7	1.5	Thickness	IR^{192}	20"	343	20	1T	B	#7
						30	1T		
8	2	Thickness	IR^{192}	20"	441	30	1T	B	#8
						40	1T		
9	0.5	Height	IR^{192}	20"	196	10	4T	A	Nil
								B	#7
10	1.1875	Height	IR^{192}	20"	270	20	2T	B	#7
11	1.5	Height	IR^{192}	20"	343	30	1T	B	#7
12	2.5	Height	IR^{192}	20"	490	45	1T	B	#8
13	0.5	Height	160	914	400	10	2T	A	Nil
								B	#6
14	1.1875	Height	170	914	600	20	1T	B	#7
15	1.5	Height	200	914	900	20	1T	B	#7
						30	1T		
16	2.5	Height	260	914	900	45	1T	B	#8

CINDE
CANADIAN INSTITUTE FOR NON-DESTRUCTIVE EVALUATION

NDT in Canada 2018 | June 19–21 | Halifax. NS

13

New Test Block

- Because of the favourable results with the MS1-maraging tool steel and Ti64-Titanium alloy test blocks, a more complex block was produced.
- A significant change was the addition of "discontinuities", of known type, dimension, orientation and position.
- Sizes for the "discontinuities" was based on the Plaque IQI dimensions of 1%, 2% and 3% of section thickness with the length for Series "A", "B" and "D" being 12.7 mm.
- Series "A", "B" and "C" were placed at 0.5 of the block step height.
- Series "C" and "D" were spherical with respect to the ultrasonic inspection. Series "D" was not flat bottomed.
- Block was produced in AlSi10Mg- Aluminum alloy.

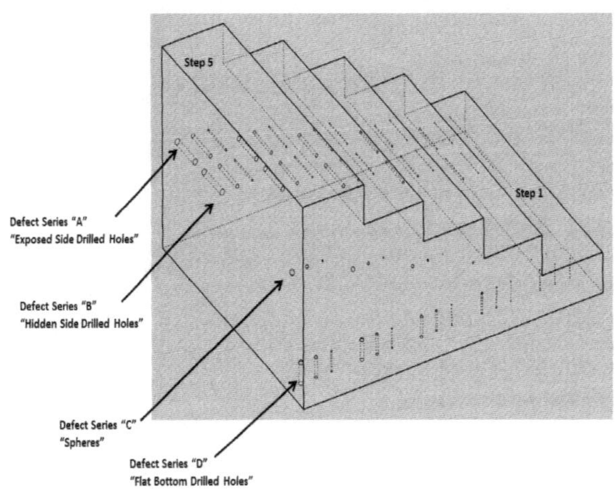

Figure 12: Step calibration block produced using AlSi10Mg

CINDE
CANADIAN INSTITUTE FOR NON-DESTRUCTIVE EVALUATION

NDT in Canada 2018 | June 19–21 | Halifax. NS

14

Radiographic Set-up for AlSi10Mg- Aluminum alloy.

- Radiographic set-up was the same for the AlSi10Mg-Aluminum alloy block as it had been for the previous blocks.
- X-ray energies only, were used and of course the IQIs were both the Plaque and Wire type.
- FFD was 914 mm.
- Type I film was used.

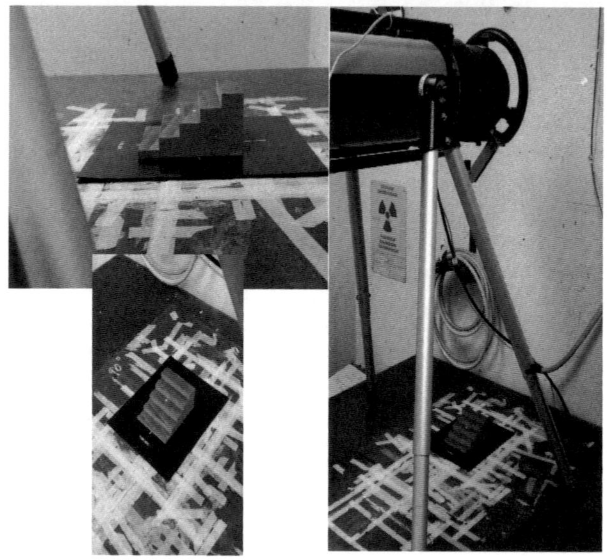

CINDE CANADIAN INSTITUTE FOR NON-DESTRUCTIVE EVALUATION

Chart 3 - Radiographic Results for AlSi10Mg- Aluminum alloy- Series A

- Defect Series "A" – Open to surface SDH – no unfused powder retention.
- With the block thickness not being what was planned, Clean-up required some material removal, the % of actual material thickness was calculated for a better understanding of the results.
- In Step 1, 2 and 3 the < 2% of Section thickness discontinuity was not detected with RT.

Aluminum Block									
Defect Series A						Note: Dimensions in mm			
Step 1		Hole Dia. (3%, 2%, 1 % of Planned Thickness)	% of Actual Thickness	Detected	Not Detected	Wire IQI	Smallest Dia. Wire Detected	Plaque IQI Designation	Hole Detected
Planned Thickness	12.7	0.381	5.44			A (6)	#3 (0.127 mm)	#5	4T (1.016 mm)
Actual Thickness	7	0.254	3.63						2T (0.508 mm)
		0.127	1.81						
Step 2		Hole Dia. (3%, 2%, 1 % of Planned Thickness)	% of Actual Thickness	Detected	Not Detected	Wire IQI	Smallest Dia. Wire Detected	Plaque IQI Designation	Hole Detected
Planned Thickness	25.4	0.762	4.01			A (6)	#4 (0.160 mm)	#17	4T (1.727 mm)
Actual Thickness	19	0.508	2.67						2T (0.864 mm)
		0.254	1.34						1T (0.432 mm)
Step 3		Hole Dia. (3%, 2%, 1 % of Planned Thickness)	% of Actual Thickness	Detected	Not Detected	Wire IQI	Smallest Dia. Wire Detected	Plaque IQI Designation	Hole Detected
Planned Thickness	38.1	1.143	3.6			B (11)	#6 (0.254 mm)	#20	4T (2.032 mm)
Actual Thickness	31.8	0.762	2.4						2T (1.016 mm)
		0.381	1.2						1T (0.508 mm)
Step 4		Hole Dia. (3%, 2%, 1 % of Planned Thickness)	% of Actual Thickness	Detected	Not Detected	Wire IQI	Smallest Dia. Wire Detected	Plaque IQI Designation	Hole Detected
Planned Thickness	50.8	1.524	3.4			B (11)	#6 (0.254 mm)	#35	4T (3.556 mm)
Actual Thickness	44.5	1.016	2.3						2T (1.778 mm)
		0.508	1.1						1T (0.889 mm)
Step 5		Hole Dia. (3%, 2%, 1 % of Planned Thickness)	% of Actual Thickness	Detected	Not Detected	Wire IQI	Smallest Dia. Wire Detected	Plaque IQI Designation	Hole Detected
Planned Thickness	63.5	1.905	3.3			B (11)	#9 (0.508 mm)	#40	4T (4.064 mm)
Actual Thickness	57.2	1.27	2.2						2T (2.032 mm)
		0.635	1.1						1T (1.016 mm)

CINDE CANADIAN INSTITUTE FOR NON-DESTRUCTIVE EVALUATION

Chart 3 - Radiographic Results for AlSi10Mg- Aluminum alloy – Series B

- Defect Series "B" – Hidden SDH – unfused powder retention.
- With the block thickness not being what was planned, Clean-up required some material removal, the % of actual material thickness was calculated for a better understanding of the results.
- With the exception of the 7 mm thickness (Step 1) all of the "SDH" were detected with RT.

Aluminum Block									
Defect Series B						Note: Dimensions in mm			
Step 1		Hole Dia. (4%, 2%, 1 % of Planned Thickness)	% of Actual Thickness	Detected	Not Detected	Wire IQI	Smallest Dia. Wire Detected	Plaque IQI Designation	Hole Detected
Planned Thickness	12.7	0.381	5.44			A (6)	#3 (0.127 mm)	#5	4T (1.016 mm)
Actual Thickness	7	0.254	3.63						2T (0.508 mm)
		0.127	1.81						
Step 2		Hole Dia. (3%, 2%, 1 % of Planned Thickness)	% of Actual Thickness	Detected	Not Detected	Wire IQI	Smallest Dia. Wire Detected	Plaque IQI Designation	Hole Detected
Planned Thickness	25.4	0.762	4.01			A (6)	#4 (0.160 mm)	#17	4T (1.727 mm)
Actual Thickness	19	0.508	2.67						2T (0.864 mm)
		0.254	1.34						1T (0.432 mm)
Step 3		Hole Dia. (3%, 2%, 1 % of Planned Thickness)	% of Actual Thickness	Detected	Not Detected	Wire IQI	Smallest Dia. Wire Detected	Plaque IQI Designation	Hole Detected
Planned Thickness	38.1	1.143	3.6			B (11)	#6 (0.254 mm)	#20	4T (2.032 mm)
Actual Thickness	31.8	0.762	2.4						2T (1.016 mm)
		0.381	1.2						1T (0.508 mm)
Step 4		Hole Dia. (3%, 2%, 1 % of Planned Thickness)	% of Actual Thickness	Detected	Not Detected	Wire IQI	Smallest Dia. Wire Detected	Plaque IQI Designation	Hole Detected
Planned Thickness	50.8	1.524	3.4			B (11)	#6 (0.254 mm)	#35	4T (3.556 mm)
Actual Thickness	44.5	1.016	2.3						2T (1.778 mm)
		0.508	1.1						1T (0.889 mm)
Step 5		Hole Dia. (3%, 2%, 1 % of Planned Thickness)	% of Actual Thickness	Detected	Not Detected	Wire IQI	Smallest Dia. Wire Detected	Plaque IQI Designation	Hole Detected
Planned Thickness	63.5	1.905	3.3			B (11)	#9 (0.508 mm)	#40	4T (4.064 mm)
Actual Thickness	57.2	1.27	2.2						2T (2.032 mm)
		0.635	1.1						1T (1.016 mm)

CINDE
CANADIAN INSTITUTE FOR NON-DESTRUCTIVE EVALUATION

NDT in Canada 2018 | June 19–21 | Halifax, NS

17

Chart 3 - Radiographic Results for AlSi10Mg- Aluminum alloy – Series C

- Defect Series "C" – Spheres – unfused powder retention.
- With the block thickness not being what was planned, Clean-up required some material removal, the % of actual material thickness was calculated for a better understanding of the results.
- At the 7 mm thickness (Step 1) the sphere at < 4% of section thickness was not detected with RT.
- At Steps 2, 3 and 5, the spheres < 1% of section thickness were not detected.
- At Step 4, it was very questionable whether the sphere was visible.

Aluminum Block									
Defect Series C						Note: Dimensions in mm			
Step 1		Hole Dia. (4%, 2%, 1 % of Planned Thickness)	% of Actual Thickness	Detected	Not Detected	Wire IQI	Smallest Dia. Wire Detected	Plaque IQI Designation	Hole Detected
Planned Thickness	12.7	0.381	5.44			A (6)	#3 (0.127 mm)	#5	4T (1.016 mm)
Actual Thickness	7	0.254	3.63						2T (0.508 mm)
		0.127	1.81						
Step 2		Hole Dia. (3%, 2%, 1 % of Planned Thickness)	% of Actual Thickness	Detected	Not Detected	Wire IQI	Smallest Dia. Wire Detected	Plaque IQI Designation	Hole Detected
Planned Thickness	25.4	0.762	4.01			A (6)	#4 (0.160 mm)	#17	4T (1.727 mm)
Actual Thickness	19	0.508	2.67						2T (0.864 mm)
		0.254	1.34						1T (0.432 mm)
Step 3		Hole Dia. (3%, 2%, 1 % of Planned Thickness)	% of Actual Thickness	Detected	Not Detected	Wire IQI	Smallest Dia. Wire Detected	Plaque IQI Designation	Hole Detected
Planned Thickness	38.1	1.143	3.6			B (11)	#6 (0.254 mm)	#20	4T (2.032 mm)
Actual Thickness	31.8	0.762	2.4						2T (1.016 mm)
		0.381	1.2						1T (0.508 mm)
Step 4		Hole Dia. (3%, 2%, 1 % of Planned Thickness)	% of Actual Thickness	Detected	Not Detected	Wire IQI	Smallest Dia. Wire Detected	Plaque IQI Designation	Hole Detected
Planned Thickness	50.8	1.524	3.4			B (11)	#6 (0.254 mm)	#35	4T (3.556 mm)
Actual Thickness	44.5	1.016	2.3						2T (1.778 mm)
		0.508	1.1	?	?				1T (0.889 mm)
Step 5		Hole Dia. (3%, 2%, 1 % of Planned Thickness)	% of Actual Thickness	Detected	Not Detected	Wire IQI	Smallest Dia. Wire Detected	Plaque IQI Designation	Hole Detected
Planned Thickness	63.5	1.905	3.3			B (11)	#9 (0.508 mm)	#40	4T (4.064 mm)
Actual Thickness	57.2	1.27	2.2						2T (2.032 mm)
		0.635	1.1						1T (1.016 mm)

CINDE
CANADIAN INSTITUTE FOR NON-DESTRUCTIVE EVALUATION

NDT in Canada 2018 | June 19–21 | Halifax, NS

18

Chart 3 - Radiographic Results for AlSi10Mg- Aluminum alloy – Series D

- Defect Series "D" – Open to surface FBH – no unfused powder retention.
- With the block thickness not being what was planned, Clean-up required some material removal, the % of actual material thickness was calculated for a better understanding of the results.
- At Steps 2 and 3 the <1% dia. hole was not visible with RT.

Aluminum Block									
Defect Series D						Note: Dimensions in mm			
Step 1		Hole Dia. (4%, 2%, 1 % of Planned Thickness)	% of Actual Thickness	Detected	Not Detected	Wire IQI	Smallest Dia. Wire Detected	Plaque IQI Designation	Hole Detected
Planned Thickness	12.7	0.381	5.44			A (6)	#3 (0.127 mm)	#5	4T (1.016 mm)
Actual Thickness	7	0.254	3.63						2T (0.508 mm)
		0.127	1.81						1T (0.254 mm)
Step 2		Hole Dia. (3%, 2%, 1 % of Planned Thickness)	% of Actual Thickness	Detected	Not Detected	Wire IQI	Smallest Dia. Wire Detected	Plaque IQI Designation	Hole Detected
Planned Thickness	25.4	0.762	4.01			A (6)	#4 (0.160 mm)	#17	4T (1.727 mm)
Actual Thickness	19	0.508	2.67						2T (0.864 mm)
		0.254	1.34						1T (0.432 mm)
Step 3		Hole Dia. (3%, 2%, 1 % of Planned Thickness)	% of Actual Thickness	Detected	Not Detected	Wire IQI	Smallest Dia. Wire Detected	Plaque IQI Designation	Hole Detected
Planned Thickness	38.1	1.143	3.6			B (11)	#6 (0.254 mm)	#20	4T (2.032 mm)
Actual Thickness	31.8	0.762	2.4						2T (1.016 mm)
		0.381	1.2						1T (0.508 mm)
Step 4		Hole Dia. (3%, 2%, 1 % of Planned Thickness)	% of Actual Thickness	Detected	Not Detected	Wire IQI	Smallest Dia. Wire Detected	Plaque IQI Designation	Hole Detected
Planned Thickness	50.8	1.524	3.4			B (11)	#6 (0.254 mm)	#35	4T (3.556 mm)
Actual Thickness	44.5	1.016	2.3						2T (1.778 mm)
		0.508	1.1						1T (0.889 mm)
Step 5		Hole Dia. (3%, 2%, 1 % of Planned Thickness)	% of Actual Thickness	Detected	Not Detected	Wire IQI	Smallest Dia. Wire Detected	Plaque IQI Designation	Hole Detected
Planned Thickness	63.5	1.905	3.3			B (11)	#9 (0.508 mm)	#40	4T (4.064 mm)
Actual Thickness	57.2	1.27	2.2						2T (2.032 mm)
		0.635	1.1						1T (1.016 mm)

CINDE
CANADIAN INSTITUTE FOR NON-DESTRUCTIVE EVALUATION

Chart 4 - Ultrasonic Results for AlSi10Mg- Aluminum alloy – Series A

- Defect Series "A" – Open to surface SDH – no unfused powder retention.
- With the block thickness not being what was planned, Clean-up required some material removal, the % of actual material thickness was calculated for a better understanding of the results.
- With the exception of the 7 mm thickness (Step 1) all of the "SDH" were detected both with UT.
- On Step 1, the UT was unable to provide the necessary resolution for the detection of the "1T and 2T" sizes which is understandable given the non-focussed nature of the probe.
- A -dB indicates signal amplitude in excess of the calibrated amplitude height.

Aluminum Block								
Defect Series A								
				RT Result		UT Result		
Inspection surface		Hole Dia. (3%, 2%, 1 % of Planned Thickness)	% of Actual Thickness	Detected	Not Detected	dB (+ or -)	Depth from Inspection Surface in/mm	
Planned Thickness	12.7	0.381	5.44			-16.1	0.089	2.2606
Actual Thickness	7	0.254	3.63					
		0.127	1.81					
Step 2		Hole Dia. (3%, 2%, 1 % of Planned Thickness)	% of Actual Thickness	Detected	Not Detected	dB (+ or -)	Depth from Inspection Surface in/mm	
Planned Thickness	25.4	0.762	4.01			5.1	0.453	11.5062
Actual Thickness	19	0.508	2.67			6.3	0.447	11.3538
		0.254	1.34			12.9	0.465	11.811
Step 3		Hole Dia. (3%, 2%, 1 % of Planned Thickness)	% of Actual Thickness	Detected	Not Detected	dB (+ or -)	Depth from Inspection Surface in/mm	
Planned Thickness	38.1	1.143	3.6			1.7	0.69	17.526
Actual Thickness	31.8	0.762	2.4			2.4	0.693	17.6022
		0.381	1.2			5.8	0.715	18.161
Step 4		Hole Dia. (3%, 2%, 1 % of Planned Thickness)	% of Actual Thickness	Detected	Not Detected	dB (+ or -)	Depth from Inspection Surface in/mm	
Planned Thickness	50.8	1.524	3.4			0.2	0.955	24.257
Actual Thickness	44.5	1.016	2.3			2.8	0.962	24.4348
		0.508	1.1			2.2	0.977	24.8158
Step 5		Hole Dia. (3%, 2%, 1 % of Planned Thickness)	% of Actual Thickness	Detected	Not Detected	dB (+ or -)	Depth from Inspection Surface in/mm	
Planned Thickness	63.5	1.905	3.3			-2.1	1.194	30.3276
Actual Thickness	57.2	1.27	2.2			-0.4	1.205	30.607
		0.635	1.1			1.5	1.236	31.3944

CINDE
CANADIAN INSTITUTE FOR NON-DESTRUCTIVE EVALUATION

Chart 4 - Ultrasonic Results for AlSi10Mg- Aluminum alloy – Series B

- Defect Series "B" – Hidden "SDH" – unfused powder retention.
- With the block thickness not being what was planned, Clean-up required some material removal, the % of actual material thickness was calculated for a better understanding of the results.
- With the exception of the 7 mm thickness (Step 1) all of the "SDH" were detected both with UT.
- On Step 1, the UT was unable to provide the necessary resolution for the detection of the "1% and 2%" sizes which is understandable given the non-focussed nature of the probe.
- A -dB indicates signal amplitude in excess of the calibrated amplitude height.

Aluminum Block								
Defect Series B								
				RT Result		UT Result		
Step 1		Hole Dia. (4%, 2%, 1 % of Planned Thickness)	% of Actual Thickness	Detected	Not Detected	dB (+ or -)	Depth from Inspection Surface in/mm	
Planned Thickness	12.7	0.381	5.44			-18	0.109	2.7686
Actual Thickness	7	0.254	3.63					
		0.127	1.81					
Step 2		Hole Dia. (3%, 2%, 1 % of Planned Thickness)	% of Actual Thickness	Detected	Not Detected	dB (+ or -)	Depth from Inspection Surface in/mm	
Planned Thickness	25.4	0.762	4.01			1.6	0.475	12.065
Actual Thickness	19	0.508	2.67			3.6	0.475	12.065
		0.254	1.34			7.3	0.492	12.4968
Step 3		Hole Dia. (3%, 2%, 1 % of Planned Thickness)	% of Actual Thickness	Detected	Not Detected	dB (+ or -)	Depth from Inspection Surface in/mm	
Planned Thickness	38.1	1.143	3.6			1	0.708	17.9832
Actual Thickness	31.8	0.762	2.4			0.8	0.71	18.034
		0.381	1.2			3.4	0.718	18.2372
Step 4		Hole Dia. (3%, 2%, 1 % of Planned Thickness)	% of Actual Thickness	Detected	Not Detected	dB (+ or -)	Depth from Inspection Surface in/mm	
Planned Thickness	50.8	1.524	3.4			1.7	0.96	24.384
Actual Thickness	44.5	1.016	2.3			2.3	0.969	24.6126
		0.508	1.1			1.9	0.982	24.9428
Step 5		Hole Dia. (3%, 2%, 1 % of Planned Thickness)	% of Actual Thickness	Detected	Not Detected	dB (+ or -)	Depth from Inspection Surface in/mm	
Planned Thickness	63.5	1.905	3.3			0.7	1.194	30.3276
Actual Thickness	57.2	1.27	2.2			0.5	1.205	30.607
		0.635	1.1			1.2	1.223	31.0642

CINDE
CANADIAN INSTITUTE FOR NON-DESTRUCTIVE EVALUATION

Chart 4 - Ultrasonic Results for AlSi10Mg- Aluminum alloy – Series C

- Defect Series "C" – "Spheres" – unfused powder retention.
- With the block thickness not being what was planned, Clean-up required some material removal, the % of actual material thickness was calculated for a better understanding of the results.
- With the exception of the 7 mm thickness (Step 1) all of the "Spheres" were detected both with UT.
- On Step 1, the UT was unable to provide the necessary resolution which is understandable given the non-focussed nature of the probe.
- To note – the significant amount of Gain for detection was possible because signal to noise ratios remained excellent.
- A -dB indicates signal amplitude in excess of the calibrated amplitude height.

Aluminum Block								
Defect Series C								
				RT Result		UT Result		
Step 1		Hole Dia. (4%, 2%, 1 % of Planned Thickness)	% of Actual Thickness	Detected	Not Detected	dB (+ or -)	Depth from Inspection Surface in/mm	
Planned Thickness	12.7	0.381	5.44			-18.4	0.109	2.7686
Actual Thickness	7	0.254	3.63					
		0.127	1.81					
Step 2		Hole Dia. (3%, 2%, 1 % of Planned Thickness)	% of Actual Thickness	Detected	Not Detected		Depth from Inspection Surface in/mm	
Planned Thickness	25.4	0.762	4.01			16	0.461	11.7094
Actual Thickness	19	0.508	2.67			16.6	0.474	12.0396
		0.254	1.34					
Step 3		Hole Dia. (3%, 2%, 1 % of Planned Thickness)	% of Actual Thickness	Detected	Not Detected		Depth from Inspection Surface in/mm	
Planned Thickness	38.1	1.143	3.6			16.6	0.683	17.3482
Actual Thickness	31.8	0.762	2.4			17.7	0.698	17.7292
		0.381	1.2			19.5	0.705	17.907
Step 4		Hole Dia. (3%, 2%, 1 % of Planned Thickness)	% of Actual Thickness	Detected	Not Detected		Depth from Inspection Surface in/mm	
Planned Thickness	50.8	1.524	3.4			20.2	0.933	23.6982
Actual Thickness	44.5	1.016	2.3			18.9	0.94	23.876
		0.508	1.1			20.1	0.964	24.4856
Step 5		Hole Dia. (3%, 2%, 1 % of Planned Thickness)	% of Actual Thickness	Detected	Not Detected		Depth from Inspection Surface in/mm	
Planned Thickness	63.5	1.905	3.3			16.2	1.166	29.6164
Actual Thickness	57.2	1.27	2.2			18.5	1.175	29.845
		0.635	1.1			20.7	1.206	30.6324

CINDE
CANADIAN INSTITUTE FOR NON-DESTRUCTIVE EVALUATION

Chart 4 - Ultrasonic Results for AlSi10Mg- Aluminum alloy – Series D

- Defect Series "D" – "FBH" – no unfused powder retention.
- With the block thickness not being what was planned, Clean-up required some material removal, the % of actual material thickness was calculated for a better understanding of the results.
- With the exception of the 7 mm thickness (Step 1) all of the "Spheres" were detected both with UT.
- On Step 1, the UT was unable to provide the necessary resolution which is understandable given the non-focussed nature of the probe.
- To note – the significant amount of Gain for detection was possible because signal to noise ratios remained excellent.
- A -dB indicates signal amplitude in excess of the calibrated amplitude height.

Aluminum Block								
Defect Series D								
				RT Result		UT Result		
Step 1		Hole Dia. (4%, 2%, 1 % of Planned Thickness)	% of Actual Thickness	Detected	Not Detected	dB (+ or -)	Depth from Inspection Surface in/mm	
Planned Thickness	12.7	0.381	5.44			-15.3	0.098	2.4892
Actual Thickness	7	0.254	3.63					
		0.127	1.81					
Step 2		Hole Dia. (3%, 2%, 1 % of Planned Thickness)	% of Actual Thickness	Detected	Not Detected		Depth from Inspection Surface in/mm	
Planned Thickness	25.4	0.762	4.01			19.3	0.544	13.8176
Actual Thickness	19	0.508	2.67			23	0.542	13.7668
		0.254	1.34			24.1	0.517	13.1318
Step 3		Hole Dia. (3%, 2%, 1 % of Planned Thickness)	% of Actual Thickness	Detected	Not Detected		Depth from Inspection Surface in/mm	
Planned Thickness	38.1	1.143	3.6					
Actual Thickness	31.8	0.762	2.4					
		0.381	1.2					
Step 4		Hole Dia. (3%, 2%, 1 % of Planned Thickness)	% of Actual Thickness	Detected	Not Detected		Depth from Inspection Surface in/mm	
Planned Thickness	50.8	1.524	3.4			15.5	1.558	39.5732
Actual Thickness	44.5	1.016	2.3			17.3	1.557	39.5478
		0.508	1.1			22	1.606	40.7924
Step 5		Hole Dia. (3%, 2%, 1 % of Planned Thickness)	% of Actual Thickness	Detected	Not Detected		Depth from Inspection Surface in/mm	
Planned Thickness	63.5	1.905	3.3			15.2	1.925	48.895
Actual Thickness	57.2	1.27	2.2			18.2	1.961	49.8094
		0.635	1.1			24.6	2.001	50.8254

CINDE
CANADIAN INSTITUTE FOR NON-DESTRUCTIVE EVALUATION

Impact of AM on Non-Destructive Evaluation(NDE)

- AM's unique part properties and part microstructures with complexity in design raises challenges for NDE.
  - The microstructure bears a remarkable resemblance to a multi-pass weldment (for good reason), so do we associate the possible discontinuities with known welding discontinuities?
  - Are all welding discontinuities as we know them possible? And if so, will we have the same limitations in inspection capability when inspecting AM parts?
  - Is there existing knowledge of failure mechanisms for AM parts which is understood and can be applied, as NDE processes, equipment, standards and training are all derived from this knowledge.
  - Pratt & Whitney Canada performed x-ray computed tomography on the latest test block which provided excellent results. From this inspection came the question of IQI's to suite the inspection process and its 3 dimensional capability.
  - Will current measures of the radiographic technique (existing IQI"s) be suitable for advanced inspection methods?

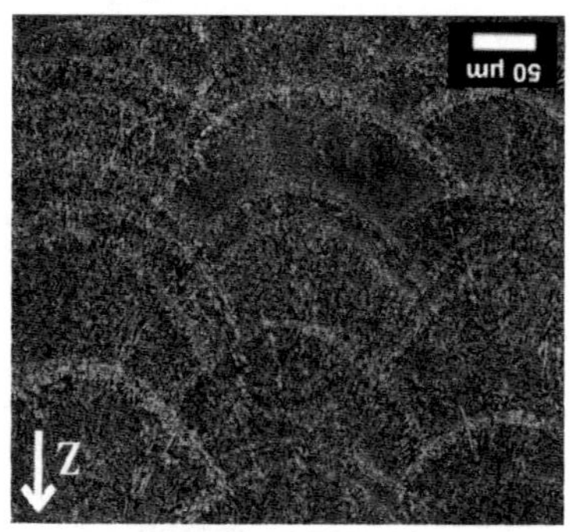

Figure 4: Optical micrographs of AlSi10Mg samples produced by means of DMLS

CINDE
CANADIAN INSTITUTE FOR NON-DESTRUCTIVE EVALUATION

Impact of AM on Non-Destructive Evaluation (NDE)

Key questions

- *How would NDE factor into such a new manufacturing process?*
  - *NDE will always be a part of the manufacturing process until a process can be proven to be 100% reliable each and every time. The cost of failing to provide a quality part for use today is much too high.*
- *Would there be any differences between traditionally manufactured components and additive manufacturing components that would influence inspection?*
  - *NDE has always been adapted to new manufacturing processes, new technologies in inspection. With the results of this work presented here, a question of redesigning the IQI for advanced inspection techniques has been initiated.*
- *What standards would guide NDE technicians?*
  - *As with all other aspects of NDE, the codes/standards and practices have continued to be adapted, redeveloped or created to suite the ever changing requirements of safety, production capability, quality and with respect to new manufacturing processes*

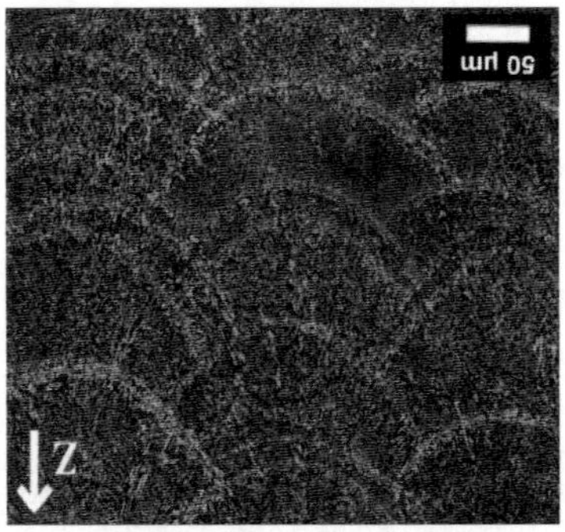

Figure 4: Optical micrographs of AlSi10Mg samples produced by means of DMLS

CINDE CANADIAN INSTITUTE FOR NON-DESTRUCTIVE EVALUATION

Redesign of the IQI for additive Manufacturing?

Design of any new IQI must consider:

- IQI must be kept near the surface of the part. Current IQI's are placed in contact with the surface of the part and do not incur additional geometric enlargement or the possibility of any penumbral shadow.
- Can the IQI's be successfully 3D printed with the required dimensional accuracy, especially if interior features are present.
- If interior features are present will it be possible to ensure complete removal of the unfused powders.
- What is the limitation of IQI thickness that can 3D printed with the features required, as standard IQI's are based on a percentage thickness as part of their design.
- Are there any restrictions in size or detail within the various materials/groups which can be printed.

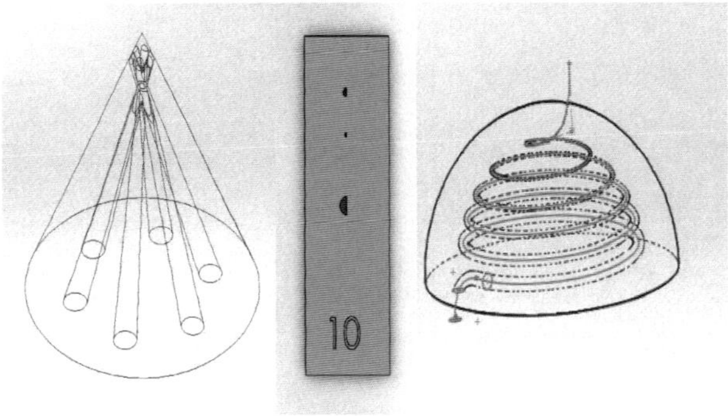

Figure 13: Possible new or redesigned IQI?

CINDE CANADIAN INSTITUTE FOR NON-DESTRUCTIVE EVALUATION

Additive Manufacturing and NDE – We will meet again

- Thanks to:
- Sharon Bond and Ciprian Pancescu of CINDE for their assistance
- Jeffrey McIsaac, General Manager AMIC,
- Mohammadarmaan Bandi, AMIC and the rest of the AMIC group for their efforts.

CINDE
CANADIAN INSTITUTE FOR NON-DESTRUCTIVE EVALUATION

NDT in Canada 2018 | June 19-21 | Halifax, NS

27

On the Relation of Residual Magnetization and Applied Stress at Stress Concentration Zone

M. Kashefi[1], T. W. Krause[2], L. Clapham[3], P.R. Underhill[2] and A.K. Krause[3]
[1]Department of Materials Science and Engineering, Ferdowsi University of Mashhad, Mashhad, Iran
[2]Department of Physics and Space Science, Royal Military College of Canada, Kingston
[3]Department of Physics Engineering Physics and Astronomy, Queen's University at Kingston,

CINDE CANADIAN INSTITUTE FOR NON-DESTRUCTIVE EVALUATION

Queen's University

NDT in Canada NDTiC 2018 Canada's NDT Conference | June 19 - 21 Halifax Convention Centre Halifax, Nova Scotia

Outline

- Introduction and Motivation
- Research Goals
- Magnetic Object Model
- Energy Minimization with Stress
- Experimental Setup
- Zero Field Measurements
- Discussion
- Summary

Queen's University

Introduction and Motivation

- Characterization of stress state is a key element in identifying the susceptibility of a material to various failure modes including:
  - Stress Corrosion Cracking
  - Potential for Plastic Deformation
  - Limited Ultimate Tensile strength
  - Fatigue Fracture Limits

Research Goals

- Develop method to identify effects of elastic stress and plastic deformation under near zero magnetization conditions on defects in steel.
- Investigate combined effects of stress and material properties such as grain size and microstructure on residual magnetization through modeling and direct laboratory measurements for validation.

Domain Configurations

Ferromagnetic Domains form in order to minimize total system energy.

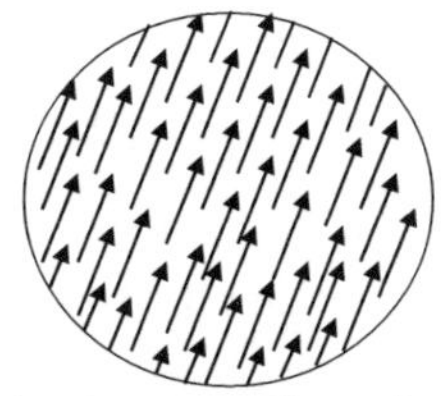

Energy terms include:

- Exchange Energy (Spin-Spin interaction)
- Magnetostatic Energy (Demagnetizing Fields)
- Crystallographic Anisotropy Energy
- Magnetoelastic Energy – Magnetostriction
- Domain Wall (DW) Energy

Shilling and Houze IEEE Trans Magn. 10, 195 (1974).

Exchange Energy

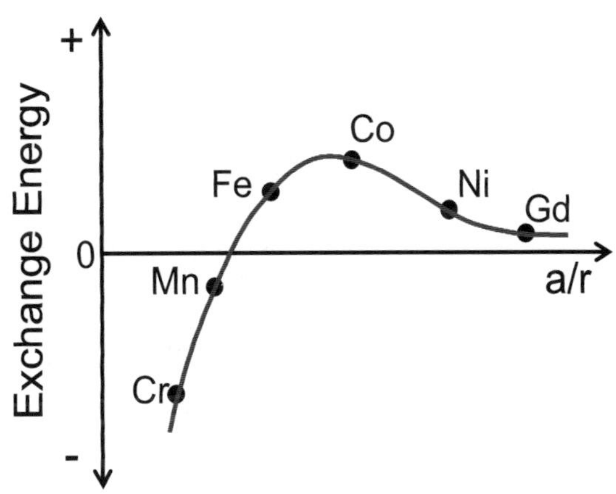

Bethe-Slater curve describes variations in exchange energy for increasing ratio of interatomic distance a radius of the 3d electron shell, r as a/r.

Magnetic Objects

- In polycrystalline steel single grains may be considered as regions exhibiting largely independent magnetic behavior (grain sizes ~ 5 to 50 μm in polycrystalline steels).
- Volumes of correlated domain behavior.
- Characterized by flux closure.
- Exhibit properties of a single Fe crystal.
- <u>Energy is minimized within a magnetic object.</u>
- How can this help us to understand magnetic behavior?

G. Berttoti IEEE Trans. Magn. **24,** 621 (1988).

Domain Wall Refinement

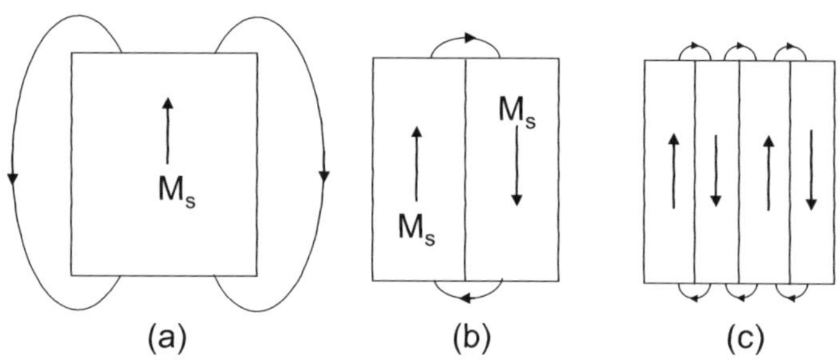

Domain wall refinement in a single ferromagnetic crystal.
a) Single domain in is spontaneously magnetized along easy axis.
b) Magnetostatic energy is reduced by splitting single domain into two domains.
c) Energy is further minimized if domains split into four.

Magnetostatic Energy Minimization

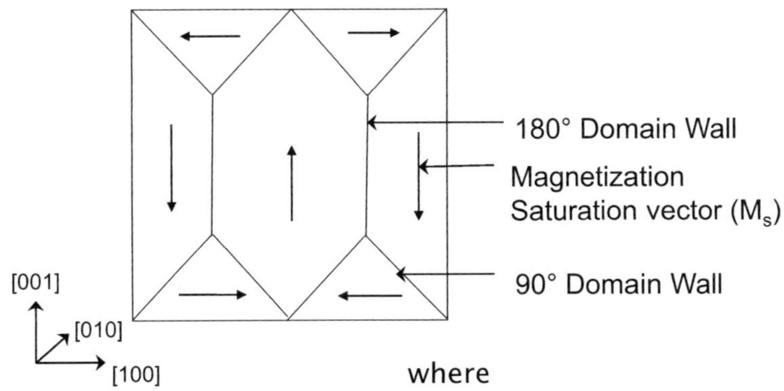

Total energy

$$E_T = E_{ex} + E_{ms} + E_K + E_{el} + E_\gamma$$

where
E_{ex} -Exchange Energy
E_{ms} -Magnetostatic Energy
E_K -Anisotropy Energy
E_{el} -Magnetoelastic Energy
E_γ - Domain Wall Energy

9

NDT in Canada 2018 | June 19-21 | Halifax, NS

Magnetoelastic Energy

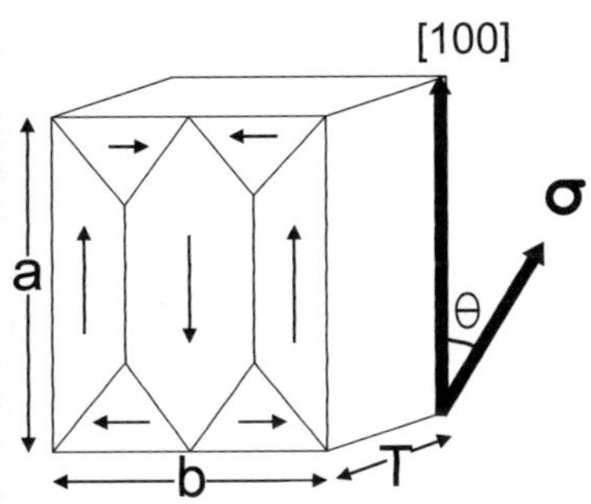

$$E_{el} = -\frac{3}{2}\lambda_{100}\sigma\left(\cos^2\theta V_{180^\circ} + \sin^2\theta V_{90^\circ}\right)$$

$$V_{90^\circ} = \frac{Tb^2}{2n}$$ Volume of 90° domains

$$V_{180^\circ} = T\left(ab - \frac{b^2}{2n}\right)$$ Volume of 180° domains

$$E_{el} = -\frac{3}{4}\lambda_{100}\sigma Tb\left(a + \left(a - \frac{b}{n}\right)\cos 2\theta\right)$$

Where:
λ_{100} -saturation strain due to magnetostriction along the [100] direction (2.07×10^{-5}).
σ - applied stress
θ – angle of σ w.r.t. crystallographic [100] direction
n – is the number of 180° domain walls

T.W. Krause, L. Clapham, A. Pattantyus and D.L. Atherton, 'Investigation of the Stress-Dependent Magnetic Easy Axis in Steel Using Magnetic Barkhausen Noise',J. Appl. Phys. **79,** 4242 (1996).

10

NDT in Canada 2018 | June 19-21 | Halifax, NS

Additional Energy Terms

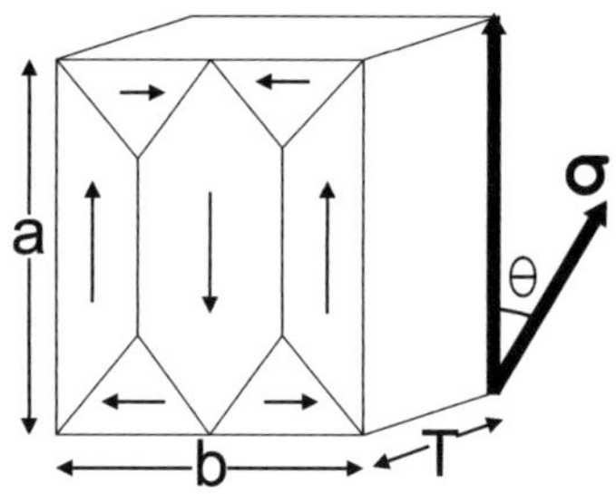

<u>Energy of 180° DWs</u>

$$E_\gamma = \gamma_{180^\circ} A_{180^\circ}$$

<u>Energy of 90° DWs</u>

$$E_\gamma = \gamma_{90^\circ} A_{90^\circ}$$

<u>Area of 180° DWs</u>

$$A_{180^\circ} = T(an - b)$$

<u>Area of 90° DWs</u>

$$A_{90^\circ} = T\sqrt{a^2 + b^2}$$

<u>Energy of Magnetic Object under uniaxial stress</u>

$$E = \frac{3}{4}\lambda_{100}\sigma Tb\left(a + \left(a - \frac{b}{n}\right)cos2\theta\right) + \gamma_{180^\circ}T(an - b) + \gamma_{90^\circ}T\sqrt{a^2 + b^2}$$

T.W. Krause and A.A. Samimi, 'Micromagnetic Techniques', Vol. 17, Nondestructive Evaluation and Quality Control, American Society of Metals, 40 pgs., In Press.

Total Energy under Formation of Magnetic Moment

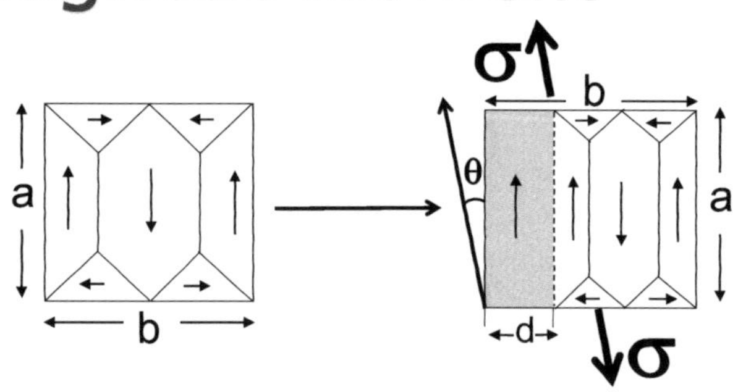

<u>Magneto-static energy per unit area (d<<a)[1]</u>

$$E_{ms} = \frac{M_s^2}{2\mu_0}a,$$

<u>Total energy becomes:</u>

$$E_T = 1.810^5 M_s^2 T d^2 - \frac{3}{4}\lambda_{100}\sigma T(b-d)\left(a + \left(a - \frac{(b-d)}{n}\right)cos2\theta\right) + \gamma_{180^\circ}T\big(an - (b-d)\big) + \gamma_{90^\circ}T\sqrt{a^2 + (b-d)^2},$$

[1]S. Chikazumi and S.H. Charap, *Physics of Magnetism*, Krieger, Florida, 1964.

Stress Dependence

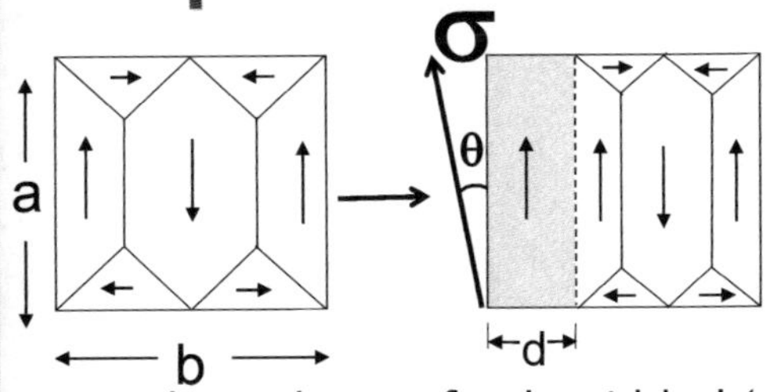

Stress dependence of pole width d (change in domain wall energy is negligible).

$$\sigma = \frac{1.4410^5 M_s^2 d^2}{\lambda_{100}(b-d)\left(a+\left(a-\frac{(b-d)}{n}\right)cos2\theta\right)}$$

Example: For a 50 μm cubic grai with n=2 180° DWs at 231 MPa

E_{el}=E_{ms}=0.624 nJ/m^3

$E_{\gamma 180^o}$=0.0044 nJ/m^3

$E_{\gamma 90^o}$= 0.0027 nJ/m^3

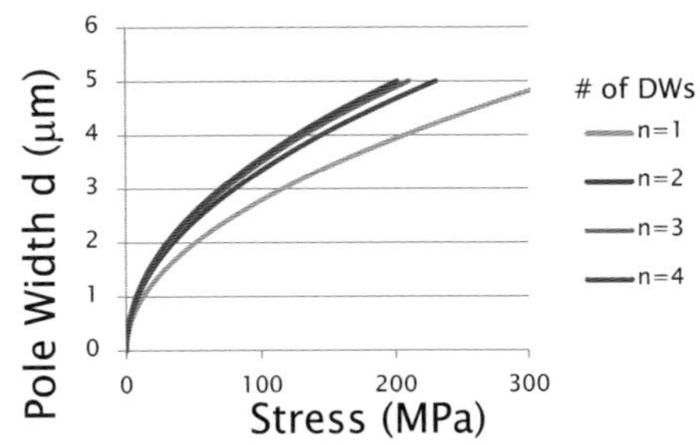

Surface Pole width versus Stress at 50 μm Grain Size

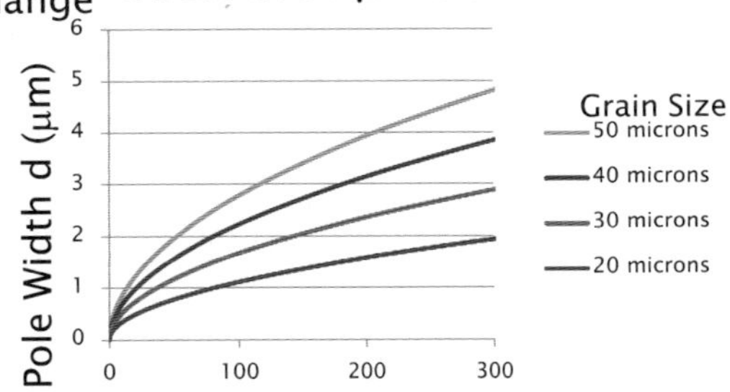

Surface Pole Width versus Stress for Various Grain sizes

13

Magnetic Dipole Moment

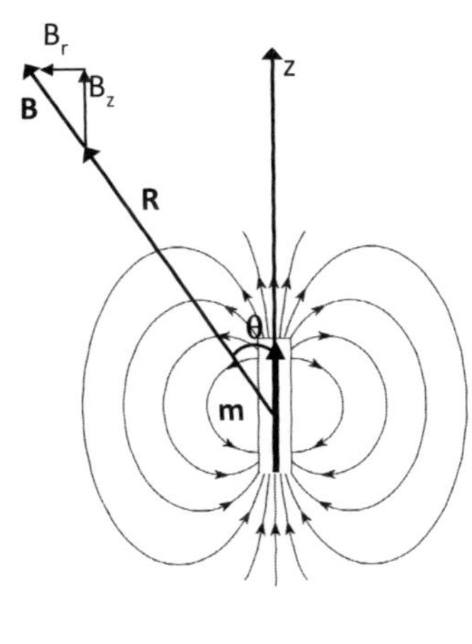

Magnetic dipole moment becomes:

$$m = (TcM_s)a,$$

which leads to magnetic field components:

$$B_r = \frac{\mu_0 m}{4\pi}\left(\frac{3\sin 2}{\theta 2R^3}\right)$$

Radial field component sensed by tangential magnetic measurements.

$$B_z = \frac{\mu_0 m}{4\pi}\left(\frac{3cos^2\theta - 1}{R^3}\right)$$

In-line magnetic field component sensed by normal magnetic component measurements.

P. Lorrain, D.R. Corson and F. Lorrain, Electromagnetic Fields and Waves, 3rd Ed., W.H. Freeman and Company, New York, 1988, P 340.

14

Experiment: Sample Parameters

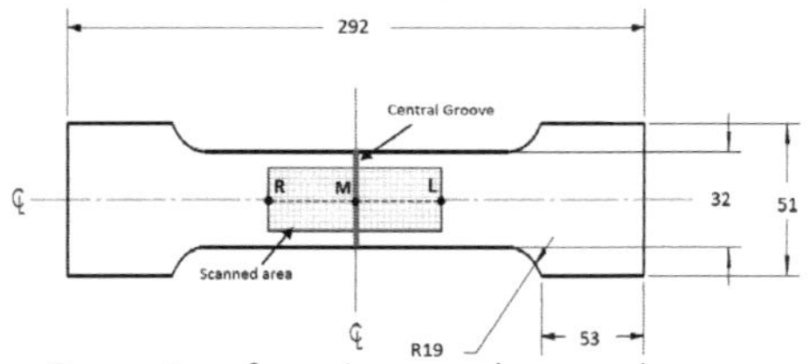

Geometry of specimen and scanned area

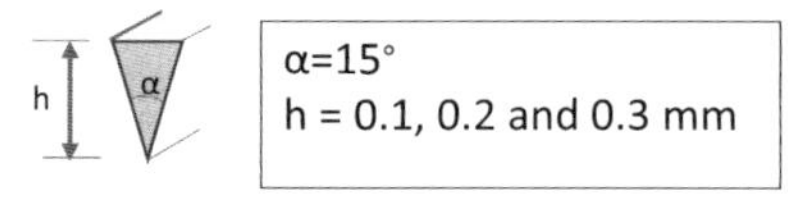

Geometry of groove

Chemical and Mechanical Properties of 1.9 mm thick 1015 Steel Sample

Chemical Composition Max.%				Mechanical Properties (MPa)		
C	Mn	P	S	σys	σuts	% ε
0.15	0.6	0.03	0.035	321	372	>25

Samples were demagnetized by AC while on tensile machine, using a U-shaped probe (Parker Research). Horizontal 20 KN tensile testing machine manufactured by Monsanto Instruments (Tensometer 20) was used for applying force with loading speed of 2 mm/min.

Scanning System

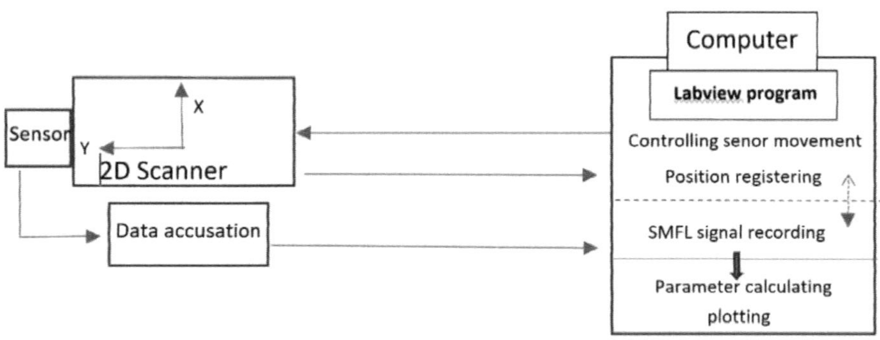

- Self magnetic flux leakage (SMFL) Scanning system .
- XY translation system is controlled by stepper motors (Arrick Robotics)
- Magnetic sensor - Honeywell 3-Axis (HMC 5883L) AMR sensor with 4.5 milliGauss (mG) resolution and ±8 Gauss range.
- 3 SMFL signal components measured over sample surface simultaneously.
- Lift-off value kept at 1 mm using a protective plastic cover on sensor.
- A I2C/SPI Interface Device (National Instruments USB-8451) was used as data acquisition device and data was recorded using a LabView program.

Signal Analysis

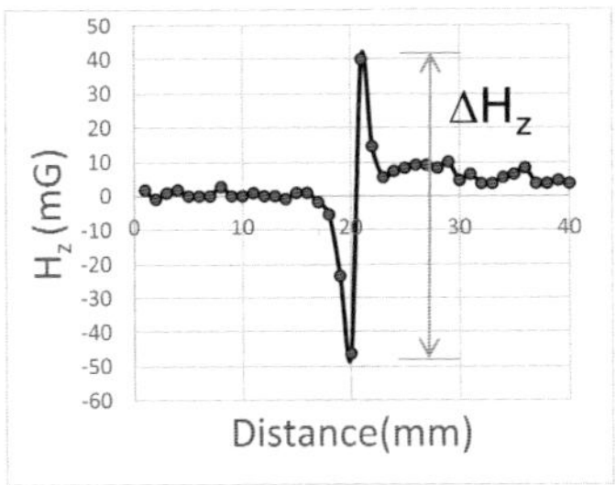

ΔH_z Definition

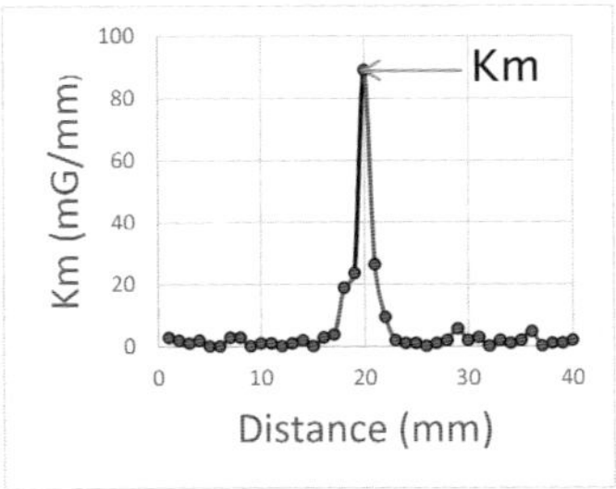

Km definition

$$Km = \frac{|dH_z|}{dx}$$

Surface plot of Km factor for sample with three grooves

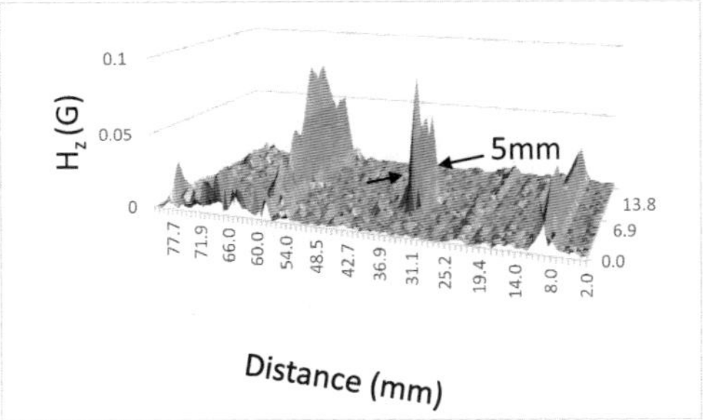

No Defect Sample Response

Changes in normal component H_z of SMFL with applied stress for un-grooved sample.

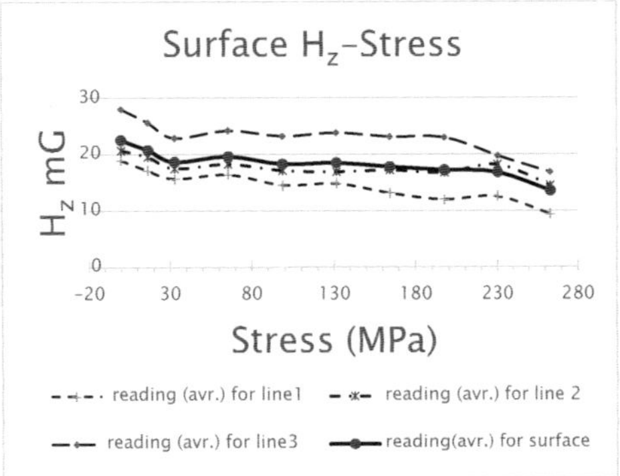

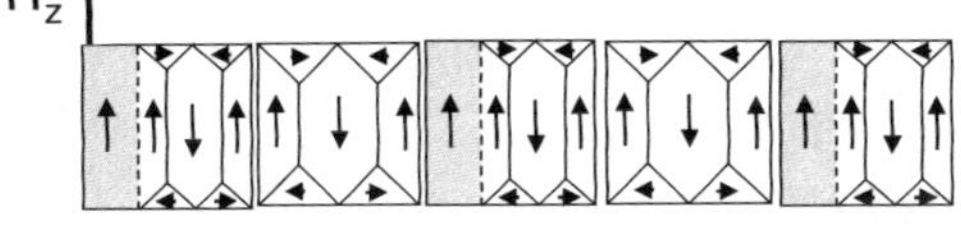

Normal H_z SMFL distribution - No Groove and No Stress

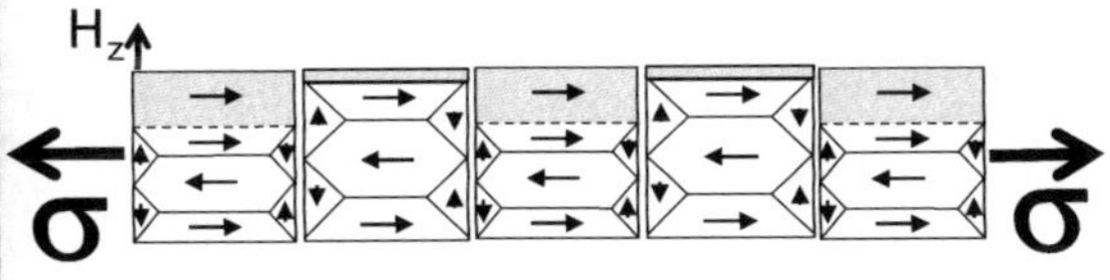

Normal H_z SMFL reduced with Stress (aligns with σ)

Surface Scan for 0.3mm Deep Groove

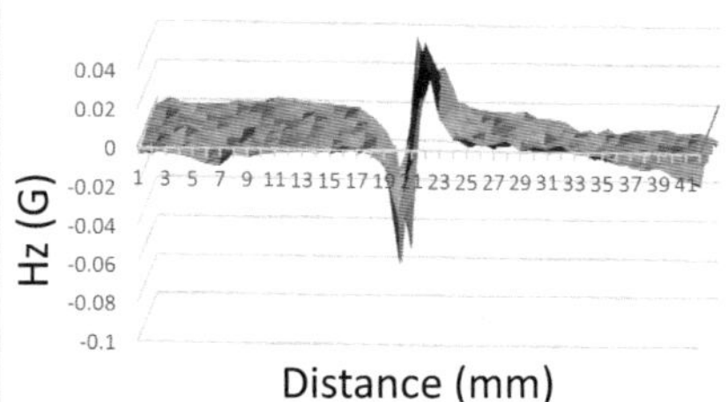

Normal H_z SMFL distribution

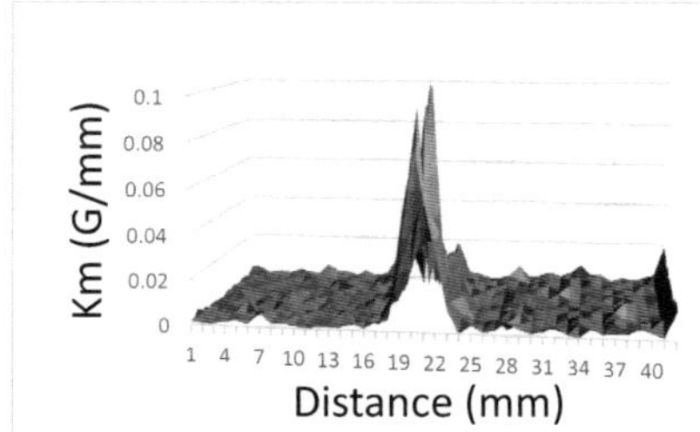

Km distribution

Change in ΔH_z with Stress over Grooves

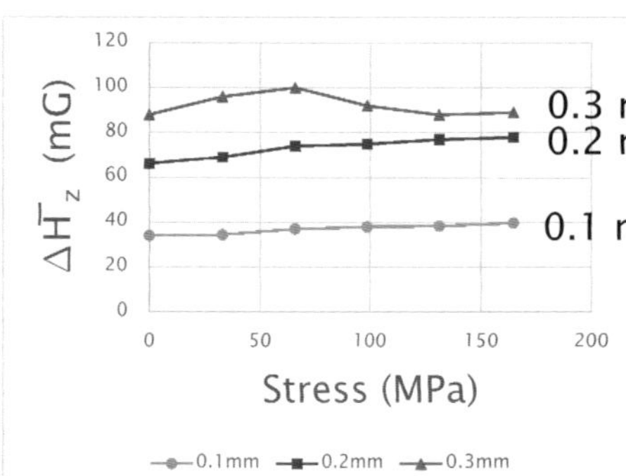

$\overline{\Delta H_z}$ with applied stress for samples with 0.1, 0.2 and 0.3 mm deep

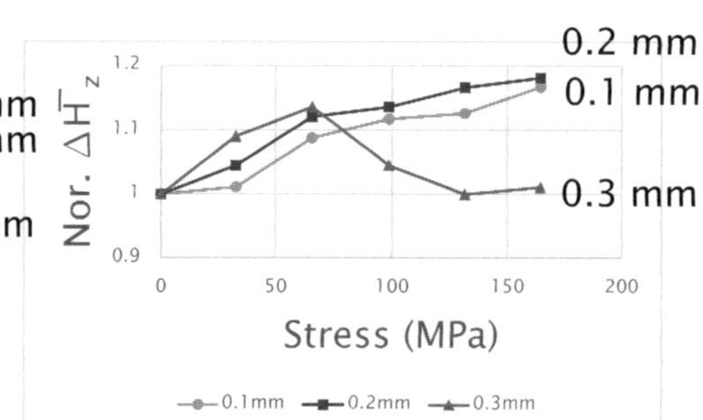

Normalized $\overline{\Delta H_z}$ with applied stress for samples with 0.1, 0.2 and 0.3 mm deep grooves.

Stress Concentration at Groove Tip

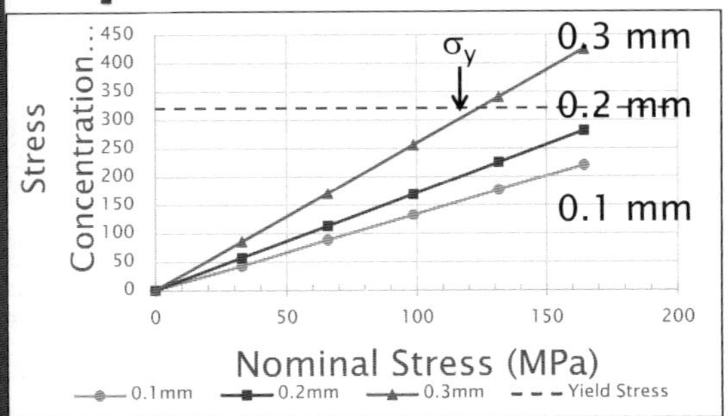

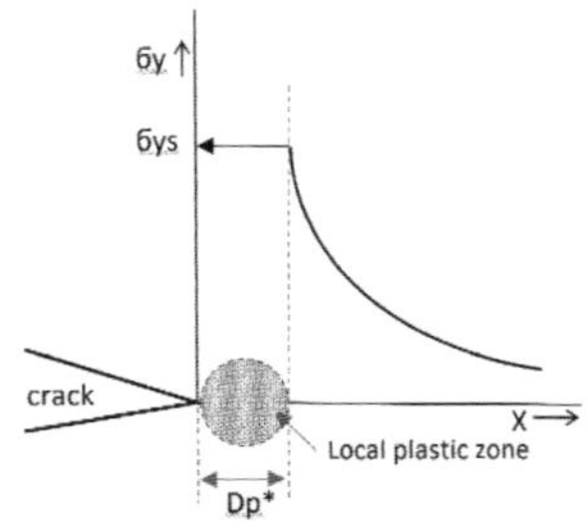

Change in stress concentration at 0.05 mm under the tip of the groove with applied stress for samples with 0.1, 0.2 and 0.3 mm deep groove.

Simplified model for formation of local plastic zone at crack tip.

21

NDT in Canada 2018 | June 19-21 | Halifax, NS

Discussion: Elastic Range

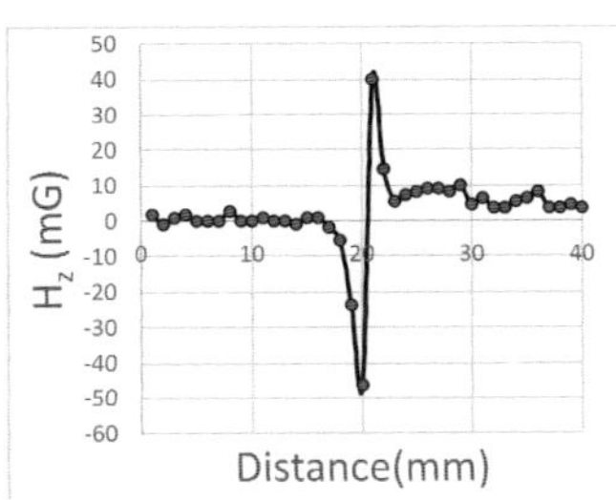

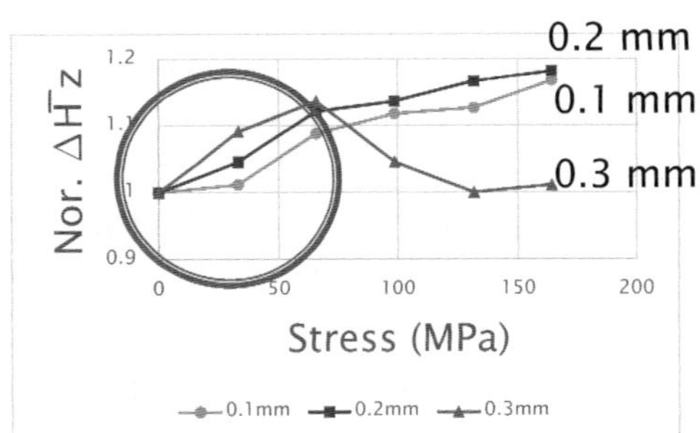

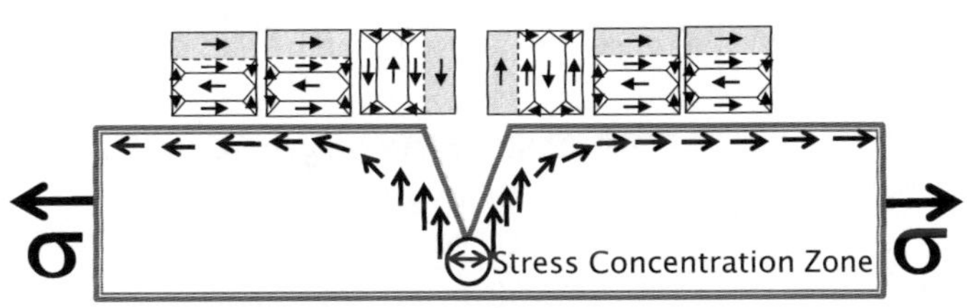

Normal H_z SMFL increases with stress concentration at defect (aligns with σ) until yield point.

22

NDT in Canada 2018 | June 19-21 | Halifax, NS

Discussion: Plastic Range

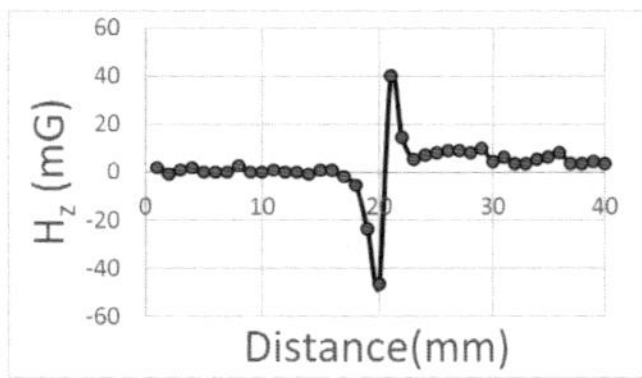

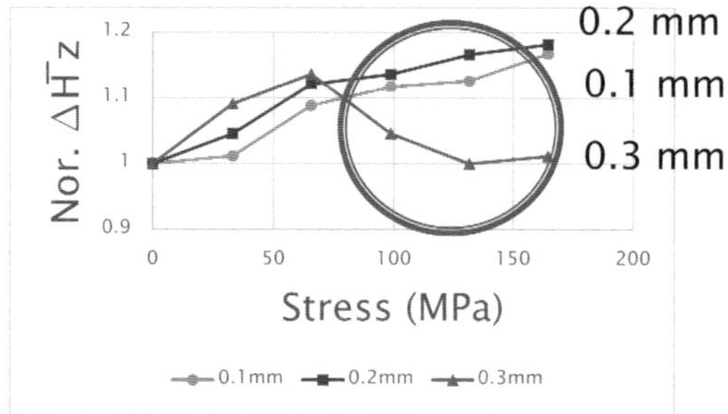

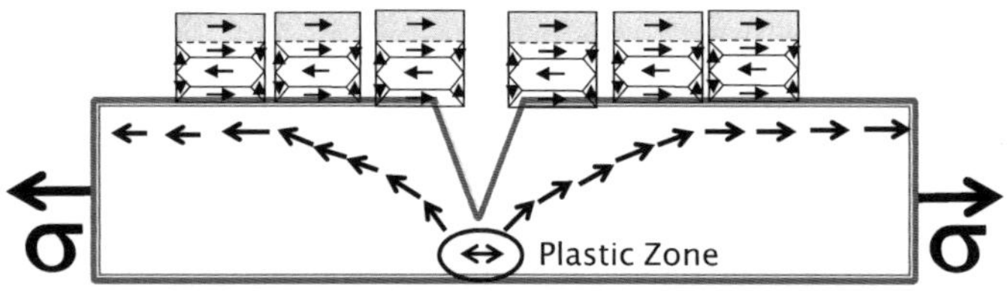

Beyond yield no additional growth in perpendicular moments occurs and moments are reoriented in direction of applied stress. Defect signal returns to its initial size.

Summary

- Work examined influence of applied stress on normal component of self magnetic flux leakage (SMFL) signals and its gradient at transverse grooves.
- SMFL signals are the result of combined effects of different elastic and plastic deformation mechanisms at the groove root, which contribute to its magnetic signature.
- Normal component of SMFL signal (ΔH_z) and its gradient (Km) increases with stress level up to local yielding.
- Local plastic deformation limits further change in magnetic properties and SMFL signals.
- Results are explained in terms conversion of magnetoelastic energy ⇔ magnetostatic surface pole energy on magnetic objects, which are detected by magnetic sensor in near zero field.

Acknowledgements

- Natural Science and Engineering Research Council Canada

Questions?

Monitoring Bridge Vibrations via Pedestrians and Mobile Sensing

Ekin OZER [1]

[1] Ekin Ozer, Middle East Technical University Northern Cyprus Campus; Mersin, Turkey
Phone: +90 535 6466494, e-mail: eozer@metu.edu.tr, eo2327@columbia.edu

Abstract
Mobile and smart sensors in structural health monitoring (SHM) applications is getting more and more widespread. A unique combination of these sensors refer to smartphone technology which poses a futuristic SHM potential if citizen engagement difficulties can be tackled. This paper presents two bridge monitoring strategies utilizing smartphone accelerometer data carried by pedestrians. These strategies utilize smartphone-based vibration measurements from walking and standing pedestrians and use it for pedestrian-induced load estimation and bridge modal identification. Comparing theoretical walk-induced forces with smartphone-based measurements of a walking pedestrian, efficiency of the former strategy is validated. The strategy associated with standing pedestrian determines vibration features of pedestrian biomechanical system, and uses it as a transfer function to isolate bridge features from the pedestrian biomechanics. With the proposed method, biomechanical effects can be eliminated from smartphone user measurements and bridge-only frequency spectra can be generated.

Keywords: structural health monitoring (SHM), mobile sensing, smartphone sensors, pedestrian bridges, biomechanical systems

1. Introduction

Understanding bridge vibrations with measurement instrumentation is one of the fundamental ways of bridge condition assessment. SHM systems developed in the last four decades show significant advances in this field [1], however, mostly require advanced equipment and extensive labour work. Mobile and smart sensing technologies emerged to solve the majority of the practical problems [2-5]. With the advent of smartphones, citizen power has become an innovative source for SHM applications [6].

Most of the smartphones are equipped with sensors such as accelerometer, gyroscope, magnetic compass, and more. Following proper crowdsourcing strategies, smartphone users can serve as SHM operators which causes no burden to system administrators and infrastructure owners [7]. If the spatiotemporal uncertainty or orientation distortion problems are resolved, it is possible to establish a smart monitoring framework for citizen engagement into the SHM process [8-9]. For example, pedestrians equipped with smartphones can measure vibrations with their phones while they are on a bridge [10].

In this paper, two major pedestrian features are investigated and two bridge monitoring solutions are proposed. The first method relies on smartphone accelerometer measurements of a moving pedestrian to estimate walk-induced forces. The second method processes standing pedestrian data to remove pedestrian biomechanical features from bridge features to illustrate bridge spectral properties. Section 2.1-2.3 introduces the theory behind the moving and stationary scenarios. Section 2.4-2.5 presents a real pedestrian and bridge experiment and the findings. Section 3 draws conclusions from this study.

2. Materials and Methods

2.1 Biomechanical Models

Biomechanical models are widely used in automotive industry, aircraft engineering, and medical studies to identify how human body interacts with vibrating environments. Adopting a similar perspective, biomechanics can be helpful to understand the relationship between pedestrians and bridges. In general, pedestrians can take numerous forms while performing any action on a bridge. Figure 1 symbolizes a number of common scenarios which can be performed by human beings.

Figure 1: Depiction of typical human activities

To mathematically and physically represent human body, numerous biomechanical models are developed by researchers. These models vary in modelling details which depends on the purpose of the model. Depending on the model detail, human body can be represented with few or many numbers of masses and springs. Figure 2 shows four different human biomechanical models with different levels of abstraction and accordingly different degree-of-freedom numbers [11-14].

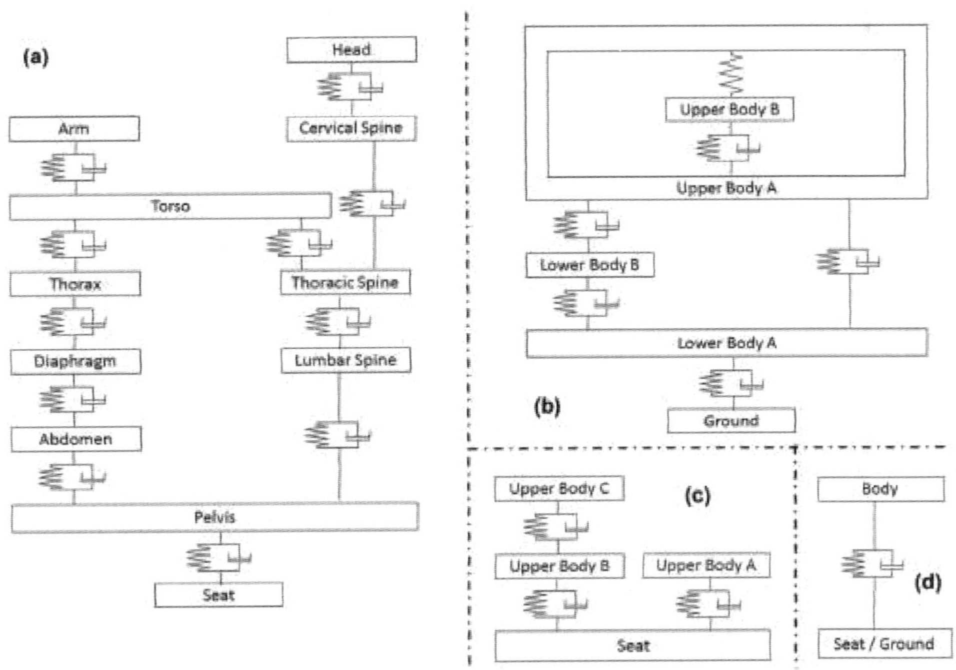

Figure 2: Exemplary biomechanical models existing in the literature [11-14]

2.2 Walk-Induced Vibrations

Considering a pedestrian as a mechanical system, dynamic forces are imposed on a bridge depending on pedestrian activity. A pedestrian moving on a bridge (e.g. walking or running) exerts harmonic forces on the bridge. Following a theoretical model given in [15] and using physical parameters such as human body weight and activity rate, these motions resemble the time history and spectral features given in Figure 3. Considering the problem in inverse direction, human body motion can be captured by smartphone accelerometer carried by the pedestrian.

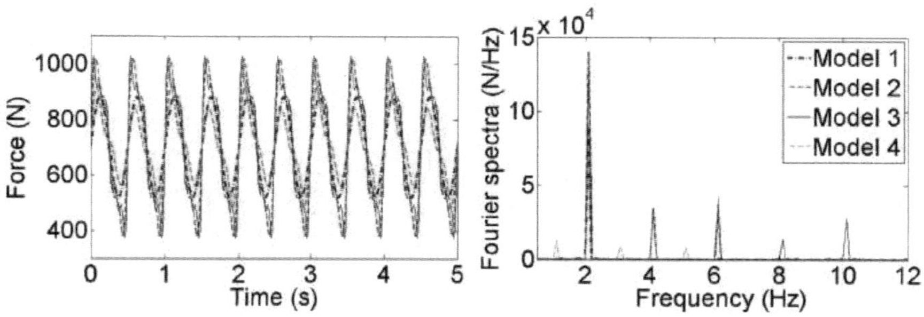

Figure 3: Time history and frequency spectra of walk-induced pedestrian forces

2.3 Standing Pedestrian Transfer Functions

Similar to walking cases, stationary pedestrian body can be represented with mathematical expressions. Considering pedestrian as a mechanical system, dynamic parameters can be determined in terms of mass, damping and stiffness. These parameters can be used to describe pedestrian body's features in the frequency domain with the help of transfer functions. Figure 4 demonstrates a variety of theoretical transfer functions representing different pedestrian bodies.

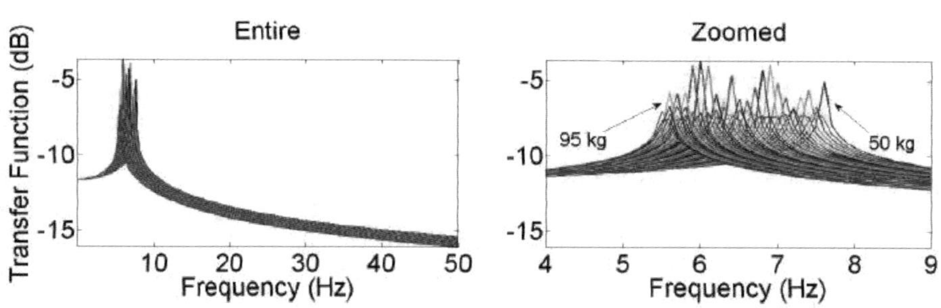

Figure 4: Theoretical transfer functions for the pedestrian

Considering standing pedestrian body as an intermediate mechanical system between the bridge and a mobile sensor, pedestrian transfer functions can be used as a tool to use indirect smartphone measurements for bridge response measurements. Sensor measurement carried by a pedestrian includes bridge vibration characteristics as well as pedestrian biomechanics. In order to isolate bridge features from pedestrian biomechanics, biomechanical system of the human body needs to be developed. Once the biomechanical system spectral features are determined, they can be removed from the smartphone data carried by the standing pedestrian can be converted into bridge-only spectral features. The entire transformation procedure of the vibration signal is depicted in Figure 5.

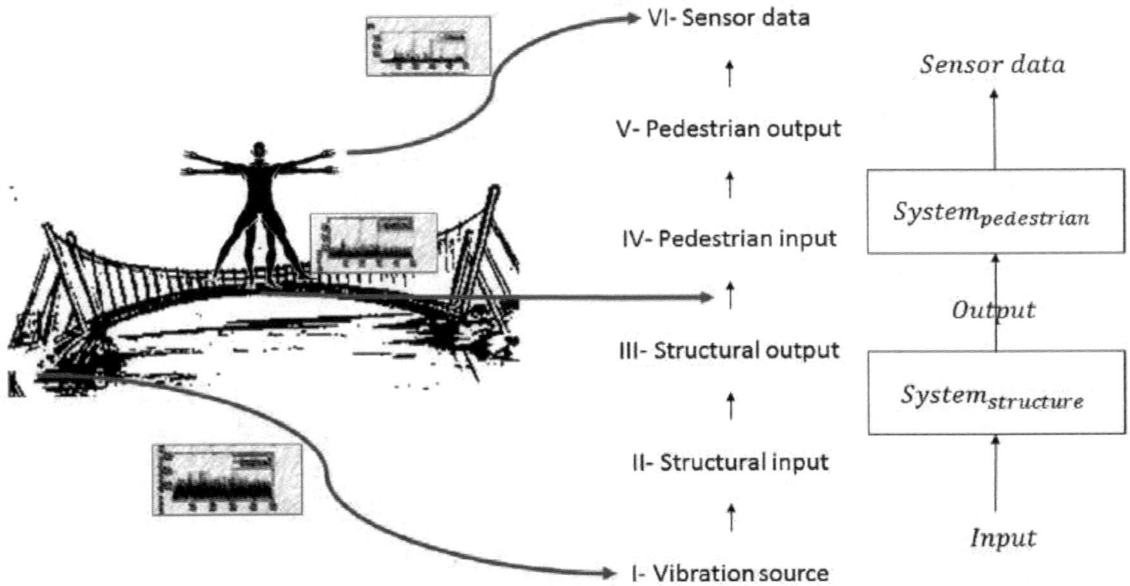

Figure 5: Transformation of the vibration signal from the source to the smartphone

2.4 Field Tests

In order to visualize the proposed methods with experimental data, eight different cases are defined. Case 1, 2, 5, 6, 7 corresponds to vibrations take from a pedestrian on a bridge, while Case 3, 4, and 8 reflects pedestrian standing on fixed ground. In Case 1 and 3, the accelerometer measurements are taken while the pedestrian is moving whereas the rest of the cases correspond to a standing pedestrian. Case 1-4 represents a smartphone carried in a backpack while Case 7 and 8 represents a smartphone carried in a pocket. Case 5 and 6 are the scenarios where there is no pedestrian intervention, and the phone is either in a backpack on the bridge, or directly coupled with the deck surface. Figure 6 shows typical conditions of the smartphone carried by the pedestrian and Figure 7 shows time histories and spectral characteristics of the smartphone measurements of each case.

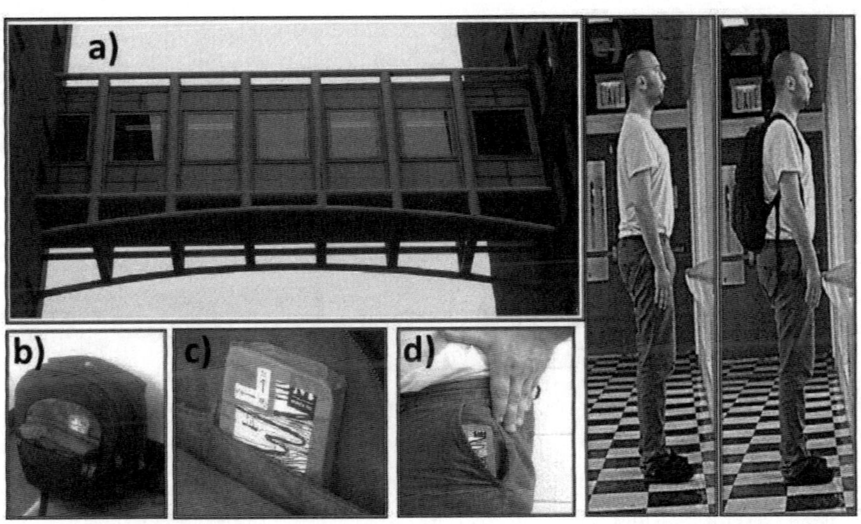

Figure 5: Test-bed bridge and smartphone configuration carried by the pedestrian

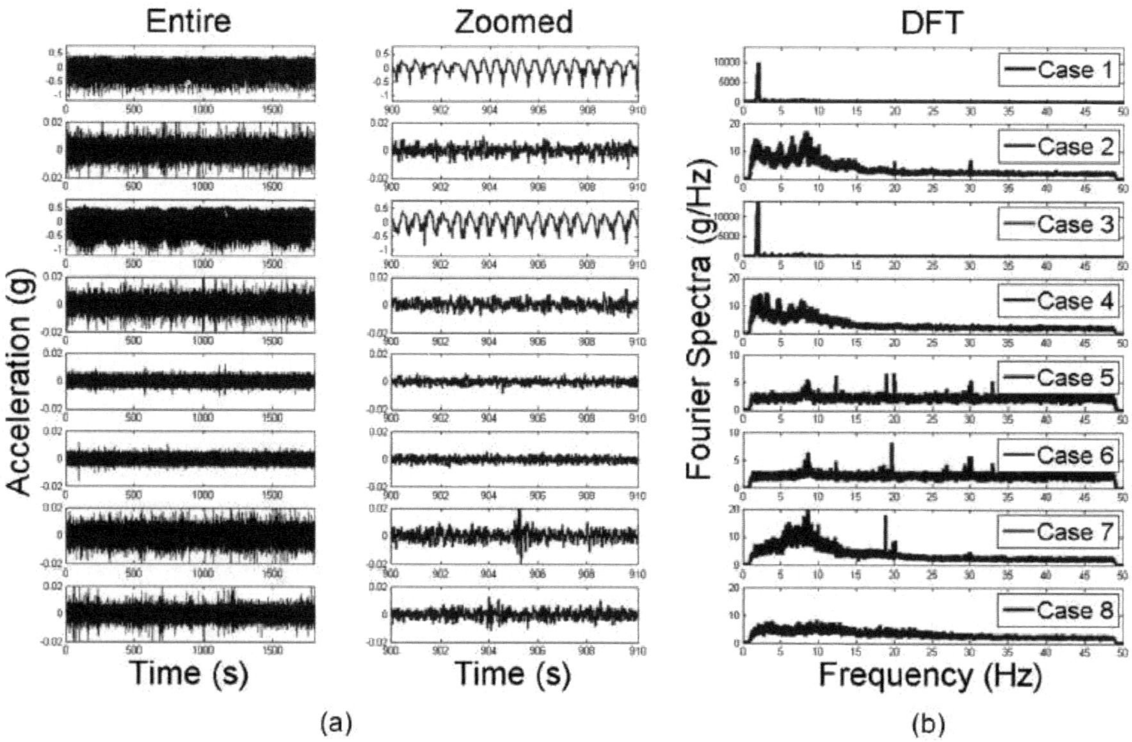

Figure 6: Time histories and spectral characteristics of pedestrian measurements

2.5 Results and Discussion

In summary, 8 cases considering different pedestrian and smartphone mobility conditions are tested. Using these cases, two main scenarios are utilized for a bridge monitoring framework. These frameworks correspond to walk-induced force estimation for the moving pedestrian scenario and modal identification isolating human biomechanical effects for the standing pedestrian scenario.

One of these cases, Case 1 corresponds to the pedestrian walking on the bridge. Modifying accelerometer data in the vertical direction with pedestrian weight, acceleration is converted into walk-induced force which is shown in Figure 7. Looking at the time history and the Fourier spectra of the estimated forces, there is a similar pattern with those obtained from the theoretical models.

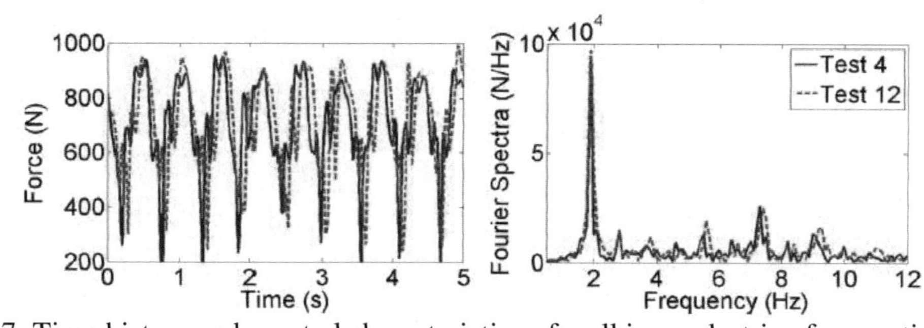

Figure 7: Time history and spectral characteristics of walking pedestrian force estimated by smartphone accelerometer

For the standing pedestrian cases, Case 2 and 7, the pedestrian stands on the bridge whereas for the Case 4 and 8, pedestrian stands on a rigid surface. Case 4 and 8 are processed to develop the biomechanical transfer function of the pedestrian. Afterwards, these transfer functions are used to remove human content from the accelerometer data carried by a standing pedestrian on the bridge. In other words, removing human biomechanical features from the indirect vibration signal, one can obtain bridge-only spectral properties and therefore, identify modal parameters. Figure 8 shows the transformation scheme and Figure 9 shows development of bridge-only spectra from the smartphone accelerometer carried by the pedestrian. Finally, Figure 10 shows results from ideal conditions where smartphone is either directly in contact with the bridge or there is no intermediate biomechanical effects even if there is no direct coupling between the sensor and the structure.

Looking at the results, it can be seen that the transformation significantly removes human-induced vibration content from the indirect signals, and represent bridge dynamics to a much better extent. However, this accuracy is higher for the high frequencies such as 20 and 30 Hz, whereas, the first mode falls within 8.5 Hz range and is still distorted by the biomechanical content up to a certain level.

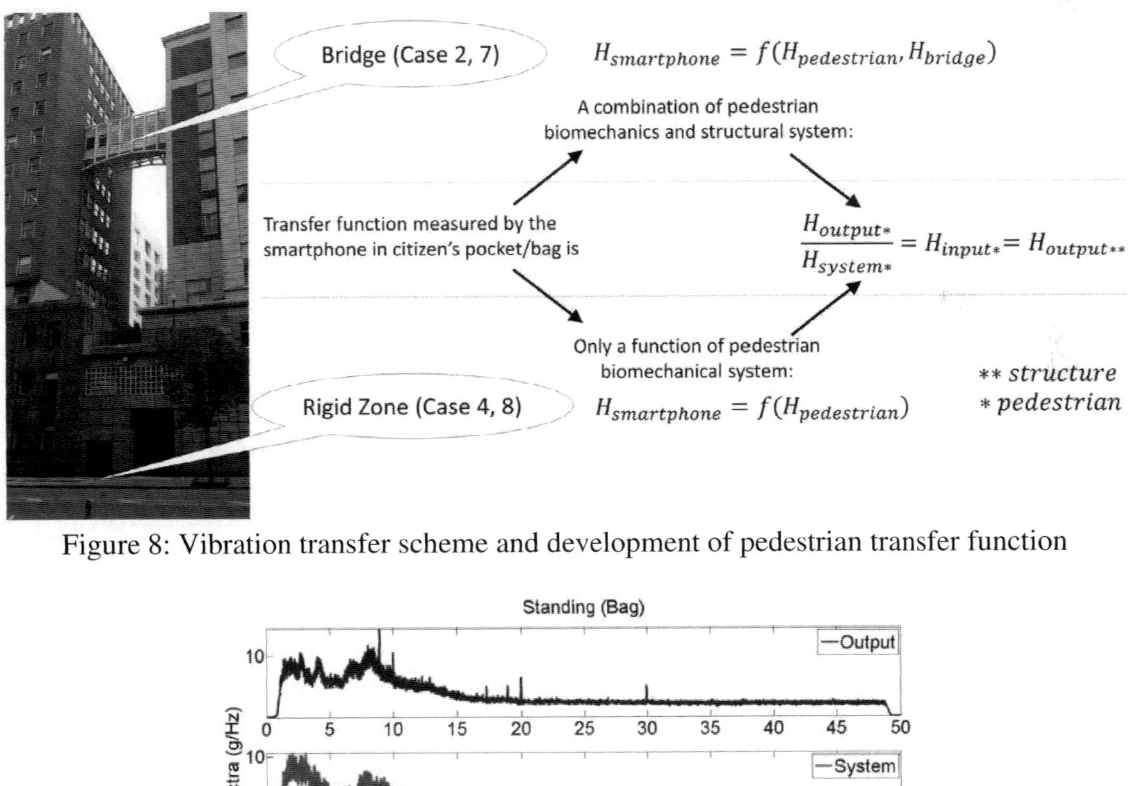

Figure 8: Vibration transfer scheme and development of pedestrian transfer function

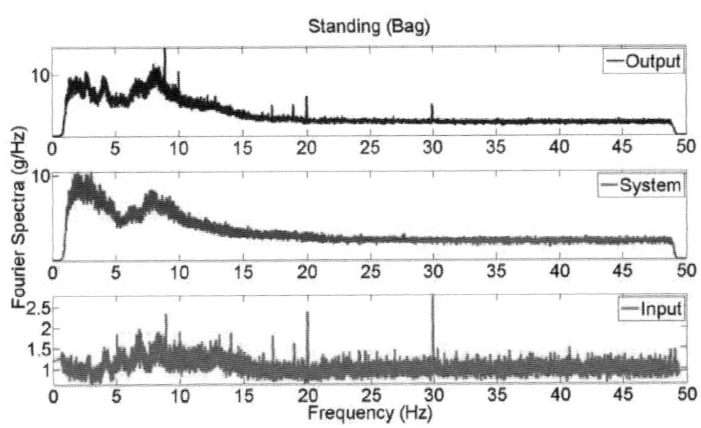

Figure 9: Conversion of smartphone-based pedestrian measurements into bridge spectral characteristics

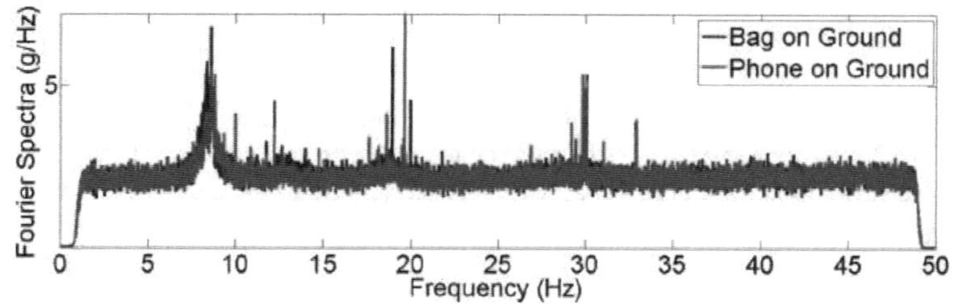

Figure 10: Reference bridge spectral characteristics without biomechanical intervention

3. Conclusion

In summary, this study presents a mobile sensing approach which utilizes pedestrian data for bridge monitoring. Using vibration data from walking pedestrians, one can estimate the dynamic load imposed on the bridge due to the human-induced activity. Using vibration data from standing pedestrians, it is possible to develop human biomechanical models with smartphone accelerometer, and use these models to eliminate biomechanical content from a sensor carried by a standing pedestrian. As a result, one can identify bridge modal characteristics using smartphone data even if there is no coupling between the sensor and the structure. The novel mobile sensing approach presented in this study can be used to understand bridge structural dynamic characteristics with a citizen-engaged and smartphone-based method. Such method can generate ubiquitous vibration datasets which is representative of city-scale bridge health monitoring applications.

References

1. Doebling, S. W., Farrar, C. R., & Prime, M. B. (1998). A summary review of vibration-based damage identification methods. Shock and vibration digest, 30(2), 91-105.
2. Taylor, S. G., Farinholt, K. M., Flynn, E. B., Figueiredo, E., Mascarenas, D. L., Moro, E. A., & Farrar, C. R. (2009). A mobile-agent-based wireless sensing network for structural monitoring applications. Measurement Science and Technology, 20(4), 045201.
3. Chen, B., & Liu, W. (2010). Mobile agent computing paradigm for building a flexible structural health monitoring sensor network. *Computer-Aided Civil and Infrastructure* Engineering, 25(7), 504-516.
4. Spencer, B. F., Ruiz-Sandoval, M. E., & Kurata, N. (2004). Smart sensing technology: opportunities and challenges. Structural Control and Health Monitoring, 11(4), 349-368.
5. Jeong, M. J., & Koh, B. H. (2009). A decentralized approach to damage localization through smart wireless sensors. Smart Structures and Systems, 5(1), 43-54.
6. Feng, M., Fukuda, Y., Mizuta, M., & Ozer, E. (2015). Citizen sensors for SHM: Use of accelerometer data from smartphones. Sensors, 15(2), 2980-2998.
7. Ozer, E., Feng, M. Q., & Feng, D. (2015). Citizen sensors for SHM: Towards a crowdsourcing platform. Sensors, 15(6), 14591-14614.
8. Ozer, E., & Feng, M. Q. (2017). Direction-sensitive smart monitoring of structures using heterogeneous smartphone sensor data and coordinate system transformation. Smart Materials and Structures, 26(4), 045026.
9. Ozer, E., & Feng, M. Q. (2016). Synthesizing spatiotemporally sparse smartphone sensor data for bridge modal identification. Smart Materials and Structures, 25(8), 085007.

10. Ozer, E., & Feng, M. Q. (2017). Biomechanically influenced mobile and participatory pedestrian data for bridge monitoring. International Journal of Distributed Sensor Networks, 13(4), 1550147717705240.
11. Mayagoitia, R. E., Nene, A. V., & Veltink, P. H. (2002). Accelerometer and rate gyroscope measurement of kinematics: an inexpensive alternative to optical motion analysis systems. Journal of biomechanics, 35(4), 537-542.
12. Curone, D., Bertolotti, G. M., Cristiani, A., Secco, E. L., & Magenes, G. (2010). A real-time and self-calibrating algorithm based on triaxial accelerometer signals for the detection of human posture and activity. IEEE transactions on information technology in biomedicine, 14(4), 1098-1105.
13. Wong, W. Y., Wong, M. S., & Lo, K. H. (2007). Clinical applications of sensors for human posture and movement analysis: a review. Prosthetics and orthotics international, 31(1), 62-75.
14. Qassem, W., Othman, M. O., & Abdul-Majeed, S. (1994). The effects of vertical and horizontal vibrations on the human body. Medical engineering & physics, 16(2), 151-161.
15. Bachmann H and Ammann W. Vibrations in structures: induced by man and machines. Zurich: International Association for Bridge and Structural Engineering, 1987.

In-service inspection; training and certification aspects

Alexander TSOUGKRANIS

TWI Ltd, Cambridge, UK
Phone: +44 (0)1223 899000, e-mail: alexander.tsougranis@twi.co.uk

Abstract
A wide range of engineering plant is subject to periodic in-service inspection in order to ensure continuous, safe and cost-effective operation by using fracture mechanics and risk-based assessments. Non-destructive testing (NDT) is an integral part of this critical assessment process and as a result, the demand for reliable in-service NDT inspections in terms of detectability and sizing accuracy is continuously increasing. The most conventional degradation mechanisms such as corrosion, stress corrosion cracking, thermal or fatigue cracking differ significantly from manufacturing flaws; their detection, quantification and accurate sizing by NDT methods require different techniques, equipment and sensitivity settings. ISO 9712 specifies in-service inspection as one of the certification sectors; however, effectiveness of the NDT inspectors on utilising the required techniques and detecting the service-induced flaws is often debatable. In this work, the cause for this ineffectiveness is being identified by looking at the standard route to NDT certification according to ISO 9712. Focusing on how and at which level the necessary sector specific competencies are embedded in the training and examination processes the conclusion reached is that in-service inspection sector is not sufficiently covered. To address this problem many certification bodies have introduced sector specific training courses as endorsement courses to existing ones, consequently increasing the number of training hours. An alternative solution proposed in this work is the introduction of ISO 9712 compliant, employer specific schemes governed by accredited third party certification bodies in order to achieve the specific competence assurance requirements, which are related to in-service inspections.

Keywords: NDT, In-service inspection, NDT Training, Certification, ISO 9712

1. Introduction

In recent years, the unification of Non Destructive Testing (NDT) training and certification requirements under ISO 9712:2012 [1] has created a more harmonised approach to training and certification and it has made the third-party certification schemes the dominant route to certification worldwide. ISO 9712:2012 [1] specifies different certification sectors and most inspectors worldwide are certified either in single product sectors (e.g. welds) or in the industrial sector "pre and in-service inspection" (multi sector) following the training and certification process mandated by the standard. Consequently, inspectors working in different industries such as oil and gas, power generation, nuclear, automotive, motorsport, conducting manufacturing or in-service inspections are trained to the same competencies.

This work is focussing on the in-service inspection sector, which is becoming increasingly important for plant and asset life assessments and on the NDT methods mostly used in this sector. These methods include conventional and advanced ultrasonic techniques, such as phased arrays and ToFD, conventional and advanced radiography and electromagnetic NDT techniques, such as eddy current array and pulsed eddy currents.

2. The training and certification challenge

Certified inspectors in the "pre and in-service sector" are eligible to carry out inspections and assumed competent to detect, characterise and size both manufacturing and service induced flaws. However, their effectiveness on conducting in-service inspections is often debatable by industry. Training is an essential factor for NDT reliability as the main source of technical knowledge and competencies. In addition, training also affects a number of other human factors, which have an impact on inspection performance. Therefore, training becomes part of the debate.

The training quality and the certification model are two factors affecting the qualification effectiveness and therefore the NDT reliability. Accredited third-party certification bodies, such as CSWIP NDT and BINDT for the UK, are responsible for governing the training and certification and maintaining the training quality. This work is focussing on the challenges that the training model for the in-service inspection sector is facing as well as proposing solutions to resolve the issues identified.

The key areas for research and the questions that need to be answered before attempting to suggest a solution to the challenge described are the following:

- What makes in-service inspection different?
- What competencies are required for those types of inspections?
- Is the training model one of the reasons for the ineffectiveness identified?
- Are all industrial sectors facing the same problem?
- How certain industries tackle the problem described?
- How the required competencies for in-service inspections differ from the ones provided by the standard ISO 9712 certification route?
- How these competencies can be mandated through the qualification process?

To answer these questions different approaches and research tools are used. Firstly, a critical review of the existing literature provides plenty of quantitative and qualitative data on NDT reliability and the effects of human and organisational factors on it. Secondly, in cooperation with key stakeholders an insight of in-service inspection procedures is available as well as information on how certain industries tackle the problem identified. Access to this information enables us to identify the key competencies required and the way to develop those competencies. Finally, the ISO 9712 route to in-service inspection certification is analysed and the competencies developed by this process are compared with the required competencies identified.

3. The In-Service Inspection characteristics

3.1 The importance of in-service inspection

The importance of in-service inspections has been identified many years ago and the nuclear industry is the leader in the field of research in NDT inspection reliability in this sector. The forward message in the work of S. R. Doctor et. al. [2] for the U.S. Nuclear Regulatory Commission (NRC) states that "the failure of in-service inspections to detect leaking cracks raised concerns regarding the effectiveness of ultrasonic testing being conducted at nuclear power plants and showed that improvements in inspection requirements were needed". In response to poor performance, NRC conducted NDT reliability studies and the nuclear industry overall has invested years of research aiming at improving in-service inspection reliability.

In addition to the use of NDT methods for flaw detection, which is one of the main NDT scopes in any sector, in-service inspection results are extensively used in failure analysis (Astrom, Sinclair and Smalley) [3], in fracture mechanics (Doctor et al.) [4], in risk-based inspection (RBI) and in fitness for service (FFS) assessment. The American Petroleum Institute (API) characteristically states that the availability of high quality UT data is often the cornerstone for FFS and RBI decisions [5]. The published work by the UK Health and Safety Executive (HSE) on plant ageing [6] also highlights the significance of NDT inspections in assessing plant integrity and affecting the decision making process for maintenance, life extension or decommissioning of ageing plants. In addition to reliable flaw detection, these assessments also require sizing accuracy, which becomes an integral part of in-service inspection reliability.

3.2 Nature of service induced flaws

It is widely accepted that flaw detectability and sizing accuracy depends on the flaw nature and the awareness of the flaw nature. The HSE: Best practice for the procurement and conduct of NDT-Part 1 [7] states that "The ultrasonic response of a defect and hence its detectability depends upon the defect characteristics. Shape, orientation, position, roughness, branching, length and through-wall size all affect the amplitude of the signal generated by a particular probe. If the inspection is not optimised for the detection of the defects of concern then the effectiveness of the inspection could be poor".

Service induced flaws are significantly different from manufacturing flaws; therefore, awareness of the nature of the expected flaws is essential factor for reliable inspection. The testing procedure, equipment selection, sensitivity, sizing techniques, recording and acceptance levels as well as inspector's competency to follow the procedure, identify the flaws and apply the right sizing techniques should all be optimised for the detection of the flaws of concern. Additionally, in regards to the comparison of manufacturing and in-service flaws the HSE: Best practice for the procurement and conduct of NDT-Part 1 [7, p.5] makes a clear statement that service-induced flaws impose additional requirements beyond those often included in manufacturing inspections.

Different industries, environments and products will develop different flaw types, however many plants share common types of defects. For example, nuclear reactors share common issues related to exposure to aggressive environments and operating conditions such as stress, irradiation, corrosive agents, vibration and fretting (IAEA) [8]. Stress corrosion cracking (SCC) is an example of service-induced flaw that is common in different industrial sectors and applications. According to HSE: Plant ageing [6], the nature of SCC is that it is branched in nature, making it difficult to detect, characterise and size with ultrasonics. A comprehensive list of the most important service-induced flaws is available in HSE: Plant ageing [6] and they are grouped in four main categories:

- Wall thinning
- Stress-driven damage, cracking and fracture
- Physical deformation
- Metallurgical / environmental damage

The same HSE report makes extensive reference to the required competencies as well as the training and certification requirements for NDT personnel and highlights the necessity of personnel competency [6, p.39].

3.3 NDT in-service inspection methods and techniques

In addition to the awareness of service-induced flaws and their characteristics, the NDT personnel conducting in-service inspections need different competencies in selecting the appropriate equipment, produce the appropriate test procedures (Level 3 personnel) and instructions, apply the appropriate sizing techniques and having in depth knowledge of the limitations of the technique applied on specific type of flaws.

In conventional surface NDT methods, the differences are usually limited to the selection of reporting levels and acceptance criteria. However, on volumetric methods such us ultrasonics and radiography the techniques used are significantly different. An insight of NDT procedures of a major nuclear generation organisation for the location and measurement of flow-accelerated corrosion (FAC) with ultrasonics as well as the weld inspection for in-service welds reveals that the sensitivity settings and the sizing techniques are much different from typical manufacturing applications. Inspection techniques might also be different. The tandem technique with ultrasonics, the use of Near Side Detection and Sizing Transducers – NSDS for SCC detection (Marr, Ginzel and Pennie) [9], the tangential radiography for detecting wall thinning and the pulsed eddy current method are all characteristic examples. Additionally, the technological advancements in Phased Array, ToFD and Eddy Current Array continuously generate new equipment and software and require special competencies by the inspector to use them effectively. Corrosion mapping, use of matrix probes, full matrix capture (FMC), two or three axis encoding, use of different scanners, analysis software, beam modelling and eddy current array (ECA) are also characteristic examples. Finally, various screening methods including traditional and non-traditional NDT techniques have emerged over recent years and they provide large areas screening of components for significant degradation. A comprehensive list of available screening techniques for in-service inspections as well as their capabilities and limitations are provided by the HSE study: Evaluation of the effectiveness of non-destructive testing screening methods for in-service inspection [10].

4. Training and certification aspects for the in-service inspection sector

4.1 The training and certification background

Over recent years, major steps have been taken for the harmonisation of personnel certification schemes worldwide and the unification of EN 473 and ISO 9712 standards under the latest ISO 9712:2012 [1] has been a milestone in that process. The third party certification schemes has become the dominant route to NDT certification worldwide. The published guide to qualification and certification of personnel for NDT by the International Committee for NDT- ICNDT [11] summarises the benefits of such schemes in the following four aspects:

- It complies with an internationally agreed ISO standard that is increasingly being adopted worldwide;
- It utilises an internationally developed training syllabus;
- Examinations (theory and practical) are provided directly by certification bodies or through authorised qualifying bodies and authorised examination centres under the control of certification bodies (many of which are linked to national NDT societies);
- It provides a harmonised standard for training, qualification and certification of NDT personnel and can be used as the base level for more specific employer-based or third party certification relevant to particular products or installations.

On the other hand employer based schemes such as SNT-TC-1A published by the American Society of Non Destructive Testing (ASNT) are also in use and mandated by the ASME code which is used worldwide, including regions in Europe. However, there are now amendments in the code allowing the use of national or international central certification programmes, to fulfil the examination requirements of the employer's written practice, which must be in accordance with SNT-TC-1A or ANSI/ASNT CP189 [11]. Moreover, ASNT is also offering the ASNT Central Certification Program (ACCP), which is a third-party certification program and meets or exceeds the requirements of CP-106:2008, the U.S. adoption of ISO 9712:2005. The ASNT website [12] states that when the latest edition of CP-106 is published the ACCP will be revised to meet or exceed the requirements of the latest ISO 9712.

4.2 The ISO 9712:2012 route to certification

The international standard ISO 9712:2012 specifies requirements for the qualification and certification of personnel who perform industrial NDT [1]. A major element of this is the definition of certification sectors as specified in the normative Annex A of the standard. The sectors specified are product sectors and industrial sectors. Additionally, the standard specifies the minimum number and type of specimens for the Level 1 and Level 2 practical examinations [1, Annex B]. The sector specification and the minimum number of examination specimens for Level 2 certification examination, which is the majority of NDT certification worldwide, are summarized in Table 1. Industrial sectors such as metal manufacturing include all or a number of product sectors (multi-sector) and pre and in-service inspection sector includes manufacturing. The practical examination requirement for the pre and in-service inspection sector as mandated by the standard is three exam samples including castings or forgings and welds.

<table>
<tr><th colspan="3">Certification sectors</th><th colspan="3">Minimum no. of exam samples for Level 2 certification</th><th colspan="3">Type of exam specimens</th></tr>
<tr><th>Product sectors</th><th colspan="2">Industrial sectors</th><th>Product sectors</th><th colspan="2">Industrial sectors</th><th>Product sectors</th><th colspan="2">Industrial sectors</th></tr>
<tr><td>Castings (c)</td><td rowspan="5">Manufacturing</td><td rowspan="5">Pre and In-service inspection</td><td>2</td><td rowspan="5">2</td><td rowspan="5">3</td><td rowspan="5">Specimens from different product sectors or character</td><td rowspan="5">Specimens from different product sectors or character</td><td rowspan="5">c/f w Castings or forgings and welds</td></tr>
<tr><td>Forgings (f)</td><td>2</td></tr>
<tr><td>Welds (w)</td><td>2</td></tr>
<tr><td>Tubes and pipes (t)</td><td>2</td></tr>
<tr><td>Wrought products (wp)</td><td>2</td></tr>
<tr><td>Composite materials (p)</td><td colspan="2"></td><td colspan="6">To be specified by the certification body</td></tr>
<tr><td rowspan="2"></td><td colspan="2">Railway maintenance</td><td rowspan="2"></td><td colspan="2">2</td><td colspan="3" rowspan="2">Specimens from different product sectors or character</td></tr>
<tr><td colspan="2">Aerospace</td><td colspan="2">2 or 3</td></tr>
</table>

Table 1

Sector specification and minimum number of exam specimens for Level 2 certification examinations according to ISO 9712:2012

Although the standard clearly states that the specification of product sectors does not preclude the development of additional sectors to satisfy national needs [1, Annex A], most accredited certification bodies follow this sector categorization in regards to the industrial sectors. It becomes apparent that there are no special examination requirements for the in-service inspection sector and the examination requirements satisfy the needs of metal manufacturing sector only.

Similarly, the training requirements and the syllabus specified for each method and sector by certification bodies are based on the technical report CEN ISO/TR 25107:2006 - Non-destructive testing — Guidelines for NDT training syllabuses [13]. According to these guidelines, for ultrasonic testing training the recommended number of hours for knowledge on various defects related to the manufacturing processes and service-induced defects related to the defined sectors is limited to 4 hours [13, par. 6.3]. In this limited time, candidates learn about all manufacturing defects in welds and potentially other sectors as well as service induced flaws. Additionally, the practical training on a generic course is based on specimens with natural or artificial manufacturing flaws for two main reasons. Firstly, the training needs to reflect the examination requirements and prepare the candidates for the examination, and secondly, service-induced flaws cannot be easily manufactured and access on specimens with those type of defects is restricted.

Interestingly, the code of conduct prepared by the British Institute of Non Destructive Testing (BINDT) for the Programme for the Assessment of Non-Destructive Testing in Industry (PANI 3 program) clearly states that "certification schemes, including PCN, are modelled on the detection of manufacturing defects and that the defects in the test samples are artificially induced and normally represent those brought about by the manufacturing process". [14, Annex10, 2008).

Concluding, the issue identified is that the theoretical knowledge related to the nature and complexity of service induced flaws, as well as the practical competencies to detect, characterize and size these flaws are not mandated by the standard route to certification. Additionally, a great percentage of candidates participating in Level 1 or Level 2 courses have very little or no experience at all on the NDT method prior to the training course. Therefore, the courses, on the limited time available, are focusing on each method's theoretical principles and on developing the basic competencies required for detection and sizing the main manufacturing flaws.

5. The response of industry to in-service inspection issues

5.1 Nuclear Industry – A different approach

The nuclear industry has identified the issue of NDT reliability for in-service inspection decades ago. The volume of published work is significant and includes quantitative and qualitative research studies, guidelines, probability of detection studies, case studies, human factors assessments, procedure assessments and studies of the impact of the organization on inspector's performance. D'Agostino, et al. [15] on the literature review of published work with key words including NDT, training, human factors or ultrasonics, states that 50 out of 91 relevant references were specific to the nuclear industry, 14 references were related to aviation or aerospace, only 1 came from the offshore oil industry and 26 were not specific to a single industry. The driver for this research programs and the associated investment is the fact that reliable in-service inspection is significantly critical for the nuclear industry.

Training is one of the important factors taken into account and studied in most of the published literature. It is assumed that more training or experience is highly beneficial to the reliability of NDT inspections. However, numerous studies have shown that the quantity of training and experience is not necessarily related to better inspection performance. The work of T.T Taylor et al. [16] on the evaluation of human reliability in ultrasonic in-service inspection for the nuclear industry concludes that NDT certification levels are not related to UT in-service inspection ability and that one of the least valuable aspects of the special training the inspectors received for the purpose of the program was the classroom training. On the other hand, the most valuable aspects of training were practice on specimens with actual cracks and comparing techniques with other personnel. Most research works in the nuclear industry have demonstrated similar results and it becomes evident that it is the training model, which is of great significance rather than the quantity or the certification level.

In response to these results, the nuclear industry has adopted a different approach to NDT training. Although NDT inspectors are being certified mostly through third-party certification schemes, the nuclear industry invests heavily on additional training which is procedure driven and focused on the in-service inspections sector for detecting SCC or other relative to the industry service induced flaws. The International Atomic Energy Agency (IAEA) has published training guidelines [17] and the Electric Power Research Institute (EPRI) has also published guidelines recommending procedure training of contractors that addresses the examination scope, the specific NDT procedures, equipment, and software that are to be used.

As a result, the nuclear industry runs in-house procedure driven training programs for its personnel and for contractors involved in NDT inspections. Those programs include awareness of the nature of sector specific flaws, practical training on real life inspection environment, the use of specimens with real service induced flaws, critical sizing and training on specific inspection techniques and equipment. The extend of those programs, as told by stakeholders, exceeds 10 weeks in duration annually. Additionally, more research has been carried out on the human, procedural and organisational factors affecting NDT reliability.

5.2 Other industrial sectors

Aerospace sector is also an industrial sector that tackles in-service inspections effectively. However, the aerospace sector is defined as different industrial sector in ISO 9712:2012 and the approach to NDT certification is generally different and outside the scope of this work. What both nuclear and aerospace industries have in common is the fact that they are both quality driven rather than cost driven.

All other, cost driven industries included in the "pre and in-service inspection" sector are lagging behind in tackling the issue identified. The studies on reliable in-service inspection in other industries is very limited and relatively recent. The UK's HSE initiated and sponsored the Programme for the Assessment of NDT in Industry (PANI) starting in 1997 and running through to 2007. The impetus for the work came out of a desire to assess the effectiveness of NDT as applied outside of the nuclear industry in the UK. All three PANI projects were focussed on manual ultrasonics; however, B. McGrath et al [18] on their work reviewing the PANI projects state that there are many lessons, which can be successfully transferred over to other NDT techniques. Unfortunately, the results of the PANI projects and the guidelines documents issued by the UK's HSE are not widely known and very few NDT organisations have made changes in respond to the issues identified, resulting in general lack of awareness.

6. The human factor in NDT reliability

The recent work of D'Agostino, et al. [15] for the U.S. Nuclear Regulatory Commission (NRC) summarises the published literature related to human factors affecting NDT reliability and states that variations in NDE reliability cannot be attributed to a single "human factor". Instead, it is likely that many factors interact with each other in a sociotechnical system to collectively impact NDE performance (Carter & McGrath, 2013; Enkvist, Edland, & Svenson, 1999; Norros, 1998 cited in D'Agostino et al. [15, p.7].

D'Agostino et al. [15] adopts the human factors categorization scheme based on Neville Moray's sociotechnical systems model. The five categories are: task characteristics, individual differences, team or group characteristics, the physical environment, and organizational factors. Although the role of training is primarily to provide the technical knowledge and competencies to carry out the inspection tasks, it is also involved in most of the five human factor categories.

In regards to task characteristics, training is affecting the pre-job preparation, the inspector's resources and the process factors. Training also has an impact in cognitive factors, specifically in memory and perception. An inspector looks for familiar patterns in the material and compares them to previously seen patterns from either previous experience or training using a combination of both long term and working memory processes (Enkvist et al., 1999, cited in D'Agostino et al. [15, p.23]. Additionally, stress and workload, expertise, level of confidence, psychological factors, performance within a team and environmental factors all have an impact on inspection performance and they are all related to training, directly or indirectly. Finally, training strategies imposed by the organisation heavily affect the inspection performance. "Hands on" training with immediate feedback is most likely to improve performance (Spanner et al.) [19]. The HSE PANI projects are also highlighting the impact of training on human factors in regards to NDT reliability.

7. Proposed solution

7.1 Respond of training organisations and certification bodies

NDT certification bodies as well as training organizations have identified the issue and in respond, they have developed application specific courses related to in-service inspections. API offers Qualification of Ultrasonic Crack Sizing Examiners that use conventional ultrasonic and Phased Array methods (QUSE and QUSE-PA). CSWIP NDT and BINDT also offer critical sizing certification courses as well as corrosion detection and mapping certification courses for phased array and conventional ultrasonics. Although, those courses can potentially enhance NDT in-service reliability they are being offered as endorsement courses to existing certificate holders resulting in cumulatively very long training durations. Given the fact that there are no shortcuts when it comes to NDT training and inspection reliability, this seems the only course of action by following this training and certification model. However, cost driven industrial sectors are reluctant to invest in long duration training especially when this type of certification is not mandated by manufacturing or in-service inspection specifications.

A different approach would be to change the training model rather than adding training courses to achieve specific competencies. ICNDT has attempted to offer an alternative to the training model by introducing performance based qualification with the introduction of ISO TS 11774 [20]. The qualification methodology described by ISO TS 11774 is based upon the candidate's ability to demonstrate capability in detecting and sizing critical discontinuities equivalent to those to be detected and sized only after the NDT procedure has been qualified [11]. The European Federation of NDT (EFNDT) in its guidelines for NDT quality system [21] summarises the disadvantages of the latter and is highlighting the fact that there is a clear overlap in scope with ISO 9712 and that there are difficulties for recognition and mutual agreement between different sector's committees. Besides, ISO TS 11774 has still the status of a technical specification and according to the ISO regulations, on the next review it will be either transformed into an ISO standard or withdrawn.

7.2 The third-party employer specific scheme

The proposed solution in this work is also offering an alternative to the certification model by combining the employer based certification schemes and third party certification. The third-party employer specific scheme proposed is based on the principles of ISO 9712, is governed by an accredited certification body but includes involvement of industry. This model is introducing employer specific inspection requirements in addition to the standard CEN ISO/TR 25107 training requirements in one training course tailored to meet the needs of specific industries. In effect, it bridges the SNT-TC-1A and the ISO 9712 models by offering the best of both worlds.

Such a model could meet the reliable inspection characteristics in regards to training as identified by the PANI projects, such as job specific training, covering the background, equipment and procedure and defect specification [14]. It could also enhance NDT reliability regarding the effect on human factors, especially if specific material characteristics, environmental factors and development of confidence levels are introduced into the training and finally meet the requirements of cost driven industries for not extensively long, competency based training courses.

On the practical side, those schemes would start by the standard training syllabus mandated by ISO 9712 and an employer's written practice based on the principles of SNT-TC-1A schemes. The written practice may include specific techniques, materials under test, sizing and reporting requirements, specific equipment, sensitivity levels or an entire procedure. The written practice may also include a training format such as working in groups or even mandate working in a specific environment.

7.3 Challenges

This model suggests joint efforts by all involved parties. This introduces challenges that need to be addressed. The challenge for industry is that it needs to be more involved in the training and certification process by providing access to materials, flawed samples, procedures, equipment and facilities as well as investing time in producing realistic written practices meeting the inspection requirements. Challenges are also introduced for the training organizations and certification bodies. They need to significantly increase their flexibility for responding to the employer specific training requirements. Trainers should be able to adopt their training style, be aware of different techniques and be dexterous with a large range of different equipment and software. Certification bodies in cooperation with the training organizations also need to become more flexible and extend the scope of their activities.

The third-party employer specific scheme proposed in this work is the respond to mitigate the issues identified for the in-service inspection industrial sector. However, the proposed training model could be used as an alternative to the standard ISO 9712 route to certification for any industrial sector. Flexibility and adoption to change are characteristics than can only be beneficial to the process. Finally, the technological advancements and the introduction of new advanced techniques with their significant range of associated equipment and software, require a new approach to the training and certification model including inherent flexibility.

8. Conclusions

The purpose of this work was to investigate the debate over the ineffectiveness of certified inspectors performing in-service inspections. At first, the in-service inspection characteristics and competencies required are identified by investigating the research programs on NDT reliability conducted mainly by the nuclear industry and the UK HSE. In-service inspection characteristics are found to be significantly different from manufacturing inspections in regards to flaw awareness and NDT techniques utilised. Then, the standard route to training and certification according to ISO 9712 is analysed and the competencies developed by this process are compared with the competencies required for in-service inspections. It is evident that this training and certification model is focussed on detection and sizing of manufacturing flaws and no additional training and examination requirements are mandated for the in-service inspection sector. Industry has identified the problem but it is only the nuclear sector that resolves it effectively. The best practices used by the nuclear industry are investigated and the reasons that other industrial sectors are lagging behind in tackling the problem are identified. Finally, a new qualification model is proposed, the third-party employer specific scheme, which combines elements from employer-based schemes and the ISO 9712 third-party certification. This model requires joint efforts by industry, training organisations and certification bodies and its successful implementation relies on addressing the challenges highlighted.

References

1. International Standards Office. ISO 9712:2012. Non-destructive testing: Qualification and certification of NDT personnel. Brussels: ISO, 2012.

2. Doctor S.R, Cumblidge S.E, Taylor T.T and Anderson M.T. The technical basis supporting ASME Code, Section XI, Appendix VIII: performance demonstration for ultrasonic examination systems. Washington, DC, 2013.

3. Astrom T., Sinclair C. and Smalley S. Use of Non-Destructive Testing (NDT) in engineering insurance. In: International Association of Engineering Insurers, IMIA conference 2002, Zurich, Switzerland, 14-18 September 2002.

4. Doctor S.R, Bates D.J, Collins H.D, Good M.S, Hartzog H.R., Heasler P.G., Mart G.A, Simonen F.A, Spanner J.C. and Taylor T.T. Integration of non-destructive examination reliability and fracture mechanics, Richland, Pacific Northwest Laboratory, 1985.

5. American Petroleum Institute (API). QUPA, Qualification of ultrasonic testing examiners (Phased Array)', 2018. [Online] Available from: http://www.api.org/products-and-services/individual-certification-programs/certifications/apiqupa#tab_about

6. UK Health and Safety Executive. Plant ageing: Management of equipment containing hazardous fluids or pressure. Norwich: HSE, 2006.

7. UK Health and Safety Executive. Best practice for the procurement and conduct of non-destructive testing, Part 1: Manual ultrasonic inspection. Norwich: HSE, 2000.

8. International Atomic Energy Agency (IAEA). Application of non-destructive testing and in-service inspection to research reactors. Vienna: IAEA, 2001.

9. Marr J., Ginzel R.and Pennie J. Hand held ultrasonic sizing of stress corrosion cracking, www.ndt.net - 3rd MENDT - Middle East Non-destructive Testing Conference & Exhibition, Bahrain, 27-30 Nov 2005.

10. UK Health and Safety Executive. Evaluation of the effectiveness of non-destructive testing screening methods for in-service inspection. Norwich: HSE, 2009.

11. International Committee for Non-destructive Testing (ICNDT). Guide to qualification and certification of personnel for NDT. Northampton: BINDT, 2012.

12. American Society for Non-destructive Testing (ASNT). ACCP Level II certification program, 2017. [Online] Available from: https://www.asnt.org/MajorSiteSections/Certification/Get_Certified/ACCP_Questions.aspx

13. International Standards Office. CEN ISO/TR 25107:2016. Non-destructive testing - Guidelines for NDT training syllabuses. Brussels: ISO, 2006.

14. UK Health and Safety Executive. Programme for the assessment of NDT in industry, PANI 3. Norwich: HSE, 2008.

15. D'Agostino A., Morrow S., Franklin C. and Hughes N. Review of human factors research in non-destructive examination. Washington, DC, 2017.

16. Taylor, T.T., Spanner Sr., J.C., Heasler, P.G., Doctor, S.R., Deffenbaugh, J.D. Evaluation of human reliability in ultrasonic in-service inspection for intergranular stress-corrosion cracks through round-robin testing. Materials Evaluation, 2005, 47 (3), pp. 338-344.

17. International Atomic Energy Agency (IAEA). Training Guidelines in non-destructive testing techniques, Vienna: IAEA: 2008.

18. McGrath B., Carter L., Bainbridge H. and Lyon R. The Lessons of PANI for Ultrasonic and NDE in General. NDT.net, 2010. [Online] Available from: http://www.ndt.net/search/docs.php3?id=11701&file=article/jrc-nde2010/papers/72.pdf

19. Spanner Sr., J.C. Human factors impact on NDE reliability, Review of progress in quantitative non-destructive evaluation, 1988.

20. International Standards Office, ISO TS 11774:2011. Non-destructive testing: Performance based qualification. Brussels: ISO, 2011.

21. European Federation for Non-destructive Testing (EFNDT). Overall NDT quality system. Northampton: EFNDT, 2012.

Bibliography

Bertovic M., Fahlbruch B. and Mueller C. Human factors perspective on the reliability of NDT in nuclear applications, Berlin, Carl Hanser Verlag. 2013.

Heasler P.G, Taylor T.T, Spanner J.C, Doctor S.R. and Deffenbaugh J.D. Ultrasonic inspection reliability for intergranular stress corrosion cracks. Washington, DC: US NRC, 1990.

McGrath B, Wheeler J. and Bainbridge H. PANI and the Role of the Written NDT Procedure. NDT.net, 2009 [Online] Available from: www.ndt.net/index.php?id=8345

UK Health and Safety Executive. Best practice for risk based inspection as a part of plant integrity management. Norwich: HSE, 2001.

UK Health and Safety Executive. Evaluation of CEN ultrasonic testing standards for in-service inspection. Norwich: HSE, 2005.

UK Health and Safety Executive. Best practice for the procurement and conduct of non-Assessment. Norwich: HSE, 2008.

World Nuclear Association. Certification of NDE Personnel: Harmonization of international code requirements, London: WNA, 2014.

DRDC

RDDC

Magnetic Barkhausen Noise Response to Temper Embrittlement of HY-80 Steel

Aroba Saleem[1], Ross Underhill[1], Nancy Herve[2], Shannon P. Farrell[2] and Thomas Krause[1]

[1] *Department of Physics and Space Science, Royal Military College of Canada*
[2] *Dockyard Laboratory (Atlantic), Defence Research and Development Canada*

CINDE CANADIAN INSTITUTE FOR NON-DESTRUCTIVE EVALUATION

NDT in Canada
NDTC 2018
Canada's NDT Conference
June 19 - 21 Halifax Convention Centre Halifax, Nova Scotia

Applications

Challenges

- Saline water – 35,000 ppm
- Water temperatures – vary widely with latitude
- Temperature gradient – water temperature decreases with depth
- Other factors – exhaust ports, welding regions

NDT in Canada 2018 | June 19-21 | Halifax, NS

3

Material requirements for submarine applications

- Corrosion resistance
- Stable physical properties
  - High strength-to-weight ratio
  - High modulus of elasticity
  - High toughness
  - Resistance to fatigue

NDT in Canada 2018 | June 19-21 | Halifax, NS

4

Problem

- Temper Embrittlement – decrease of impact toughness
- Intergranular failure
- Occurs during post fabrication heat treatment
- Change of ductile-to-brittle transition temperature
  - –Ductile-to-brittle transition temperature of HY-80 is –18 °C
- Measurement of temper embrittlement – destructive testing

Failure due to brittle fracture

–Freezing temperature
–Impact loading
–Sulphur content
–Brittle fracture

Magnetic Barkhausen Noise Analysis

- Discontinuous magnetization changes – changing applied magnetic fields
  - Abrupt magnetic domain wall motion
- Sensitive to microstructural variations and stress state of material
  - Grain size, texture, inclusions
  - Strength and hardness
- Potential non-destructive testing (NDT) method

Magnetic Barkhausen Noise Analysis

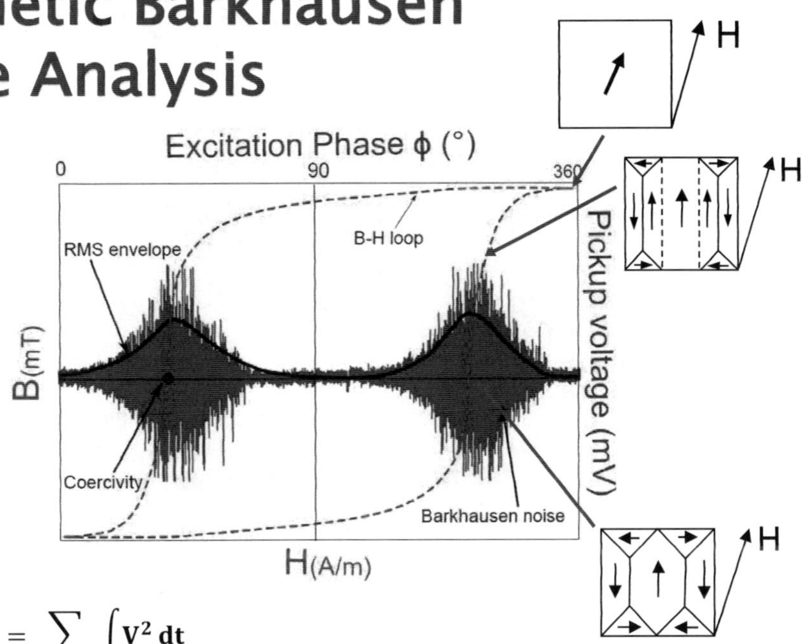

$$\mathbf{MBN_{Energy}} = \sum_{\mathbf{events}} \int \mathbf{V^2\,dt}$$

Krause et al. , Micromagnetic Techniques, ASM Handbook, 2018

Concept map

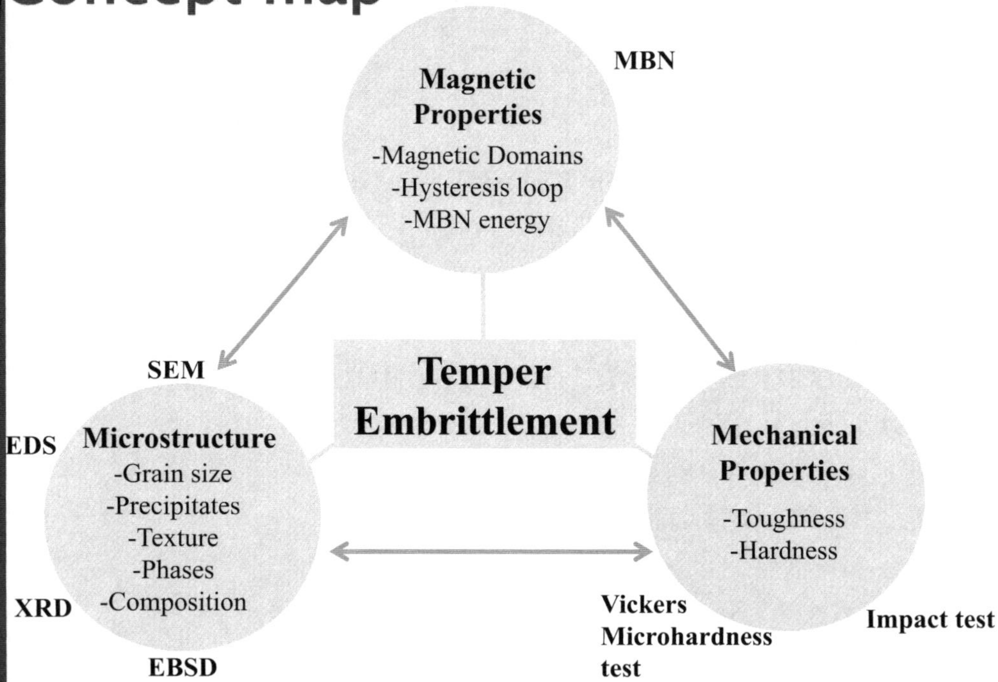

Materials and Methodology

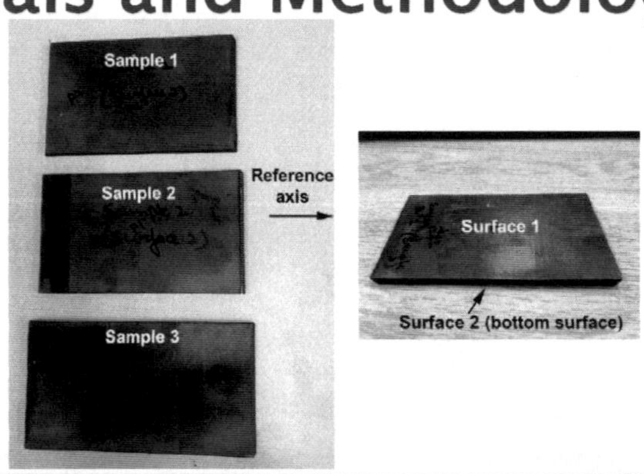

	Dimensions	Holding time at 525° C
Sample 1	114. 7 mm × 69.3 mm × 5.2 mm	48 hrs
Sample 2	120.4 mm × 69.2 mm × 5.2 mm	168 hrs
Sample 3	129.9 mm × 69.3 mm × 5.3 mm	336 hrs

Materials and Methodology

Magnetic Barkhausen Noise (MBN) Measurement

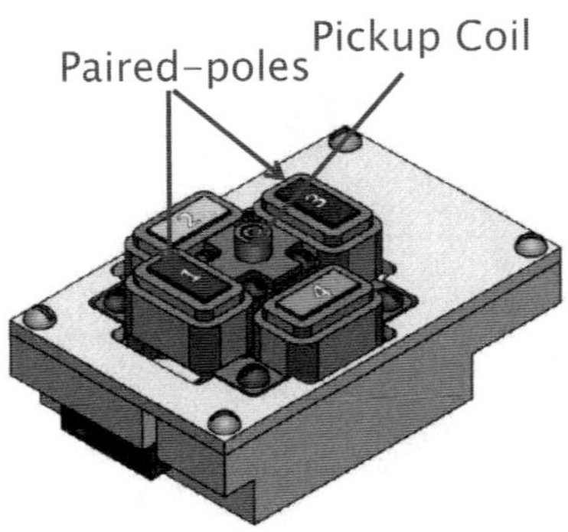

- Tetrapole probe
- Flux controlled (350 mT)
- 50 Hz

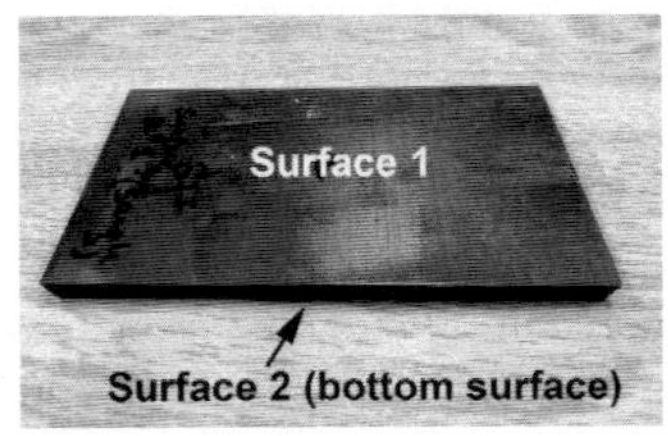

S. A. White, "A Barkhausen Noise Testing System for CANDU Feeder Pipes", Queen's University,, 2018

Materials and Methodology

Impact testing

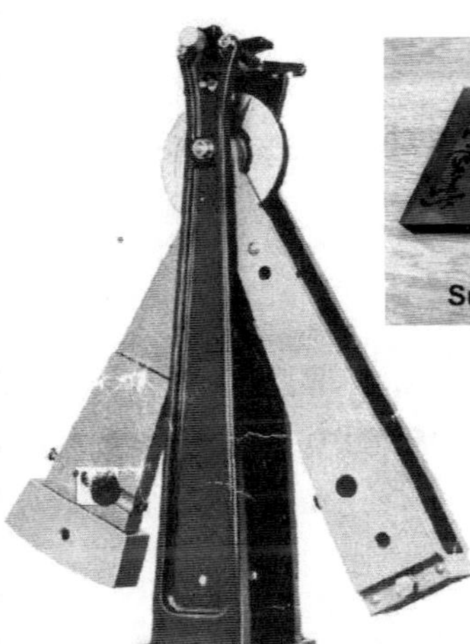

Impact tester

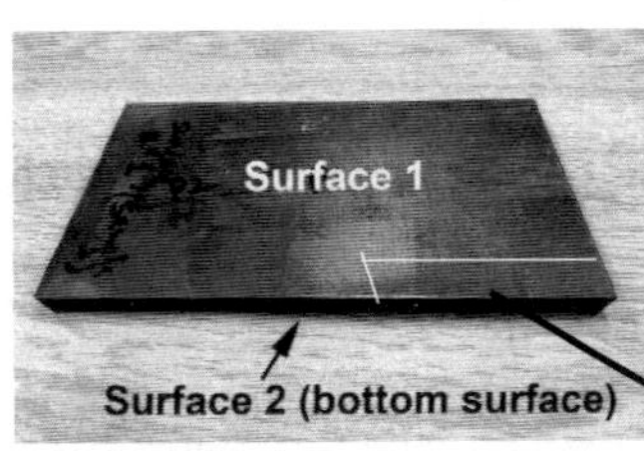

Energy absorbed until fracture (Toughness)

Sample before testing

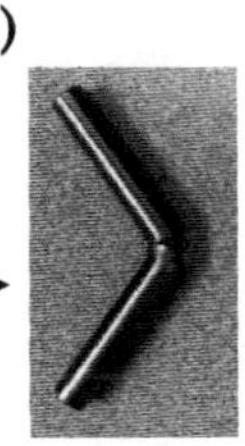

Sample after fracture

Technical Bulletin No. 202-28-14(E) M-E-1 (11/74), Balanced Impact Machine for Metals, Monsanto

Materials and Methodology

Hardness testing

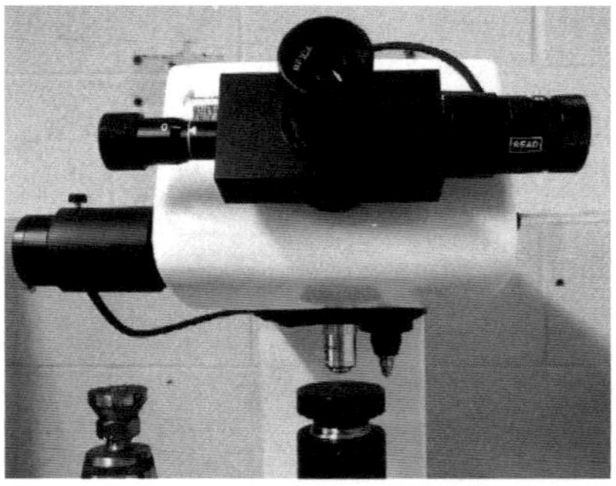

Vickers microhardness tester

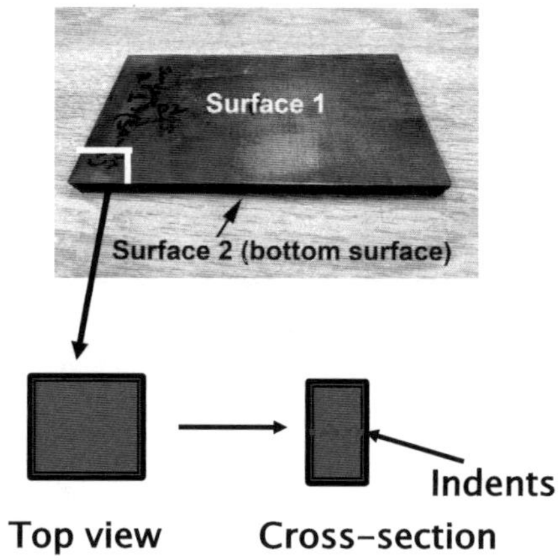

Concept map

Magnetic Properties
-Magnetic Domains
-Hysteresis loop
-MBN energy

MBN

SEM

EDS

Microstructure
-Grain size
-Precipitates
-Texture
-Phases
-Composition

XRD

EBSD

Temper Embrittlement

Mechanical Properties
-Toughness
-Hardness

Vickers Microhardness test

Impact test

MBN Results - Sample 1

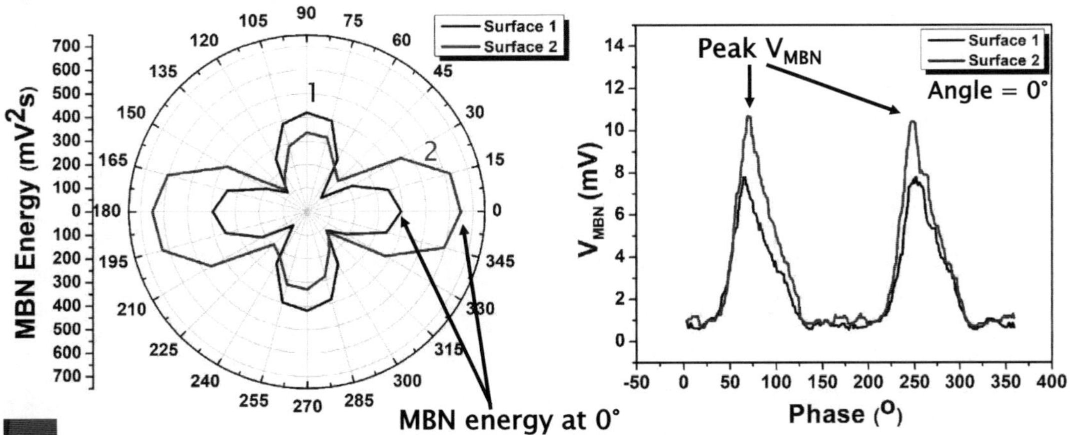

MBN Results - Sample 2

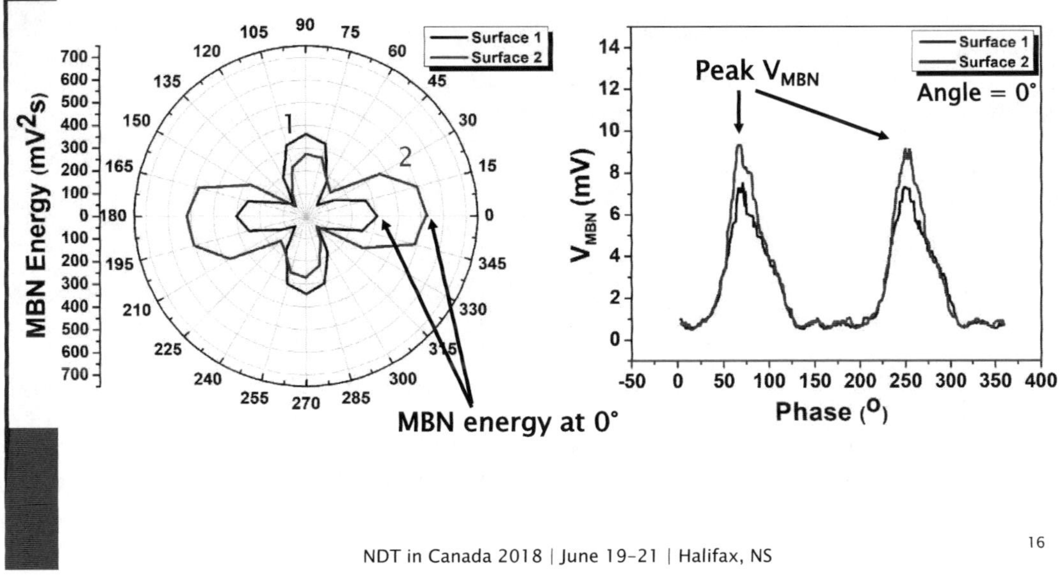

MBN Results - Sample 3

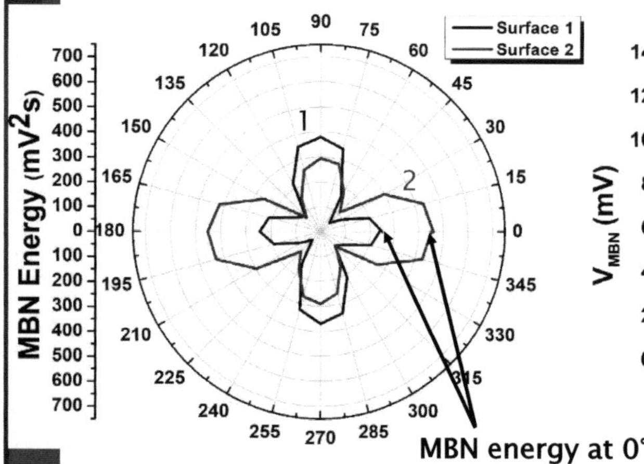

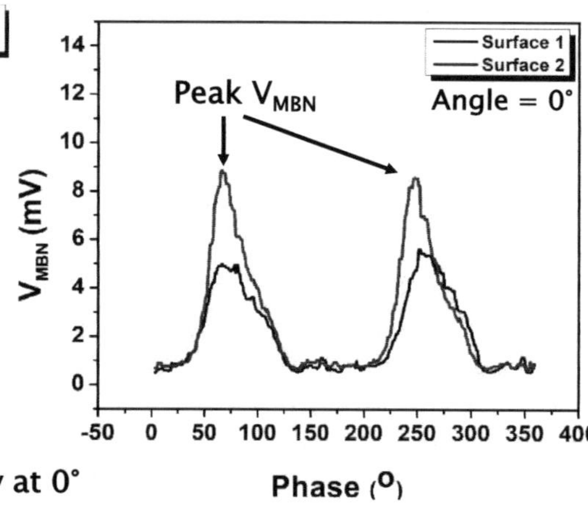

Effect of holding time

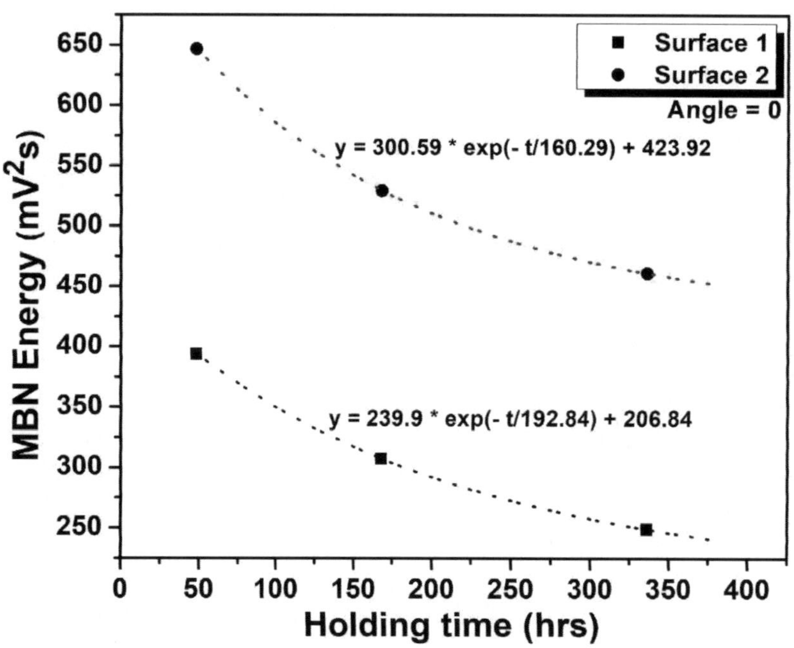

Concept map

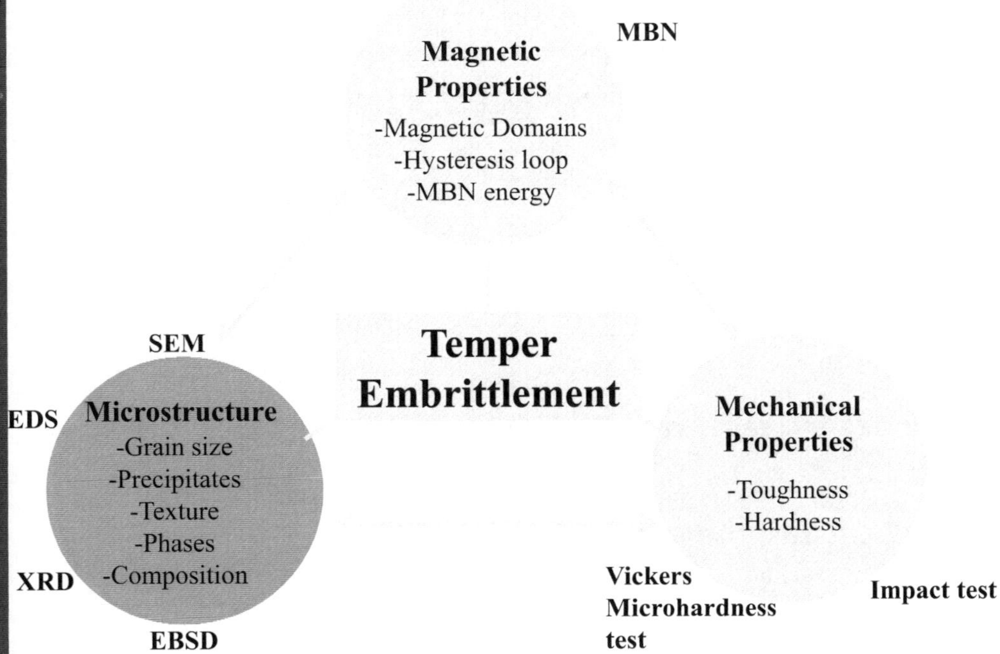

SEM Microstructure

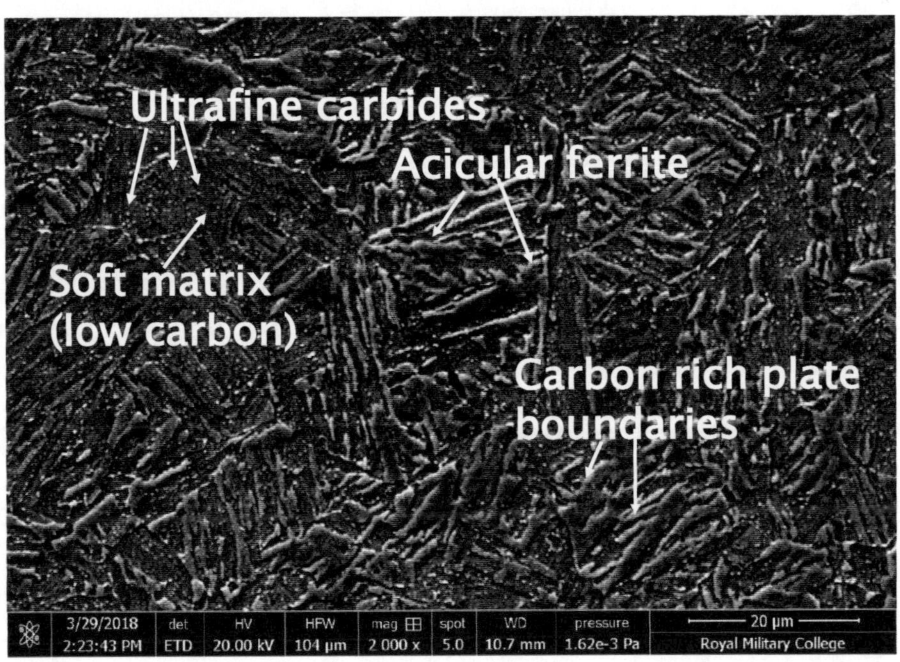

SEM Microstructure

	Soft Ferrite Matrix (μm)	Acicular ferrite (μm)
Sample 1	9 ± 2	1.9 ± 0.46
Sample 2	17 ± 3	2.68 ± 0.64
Sample 3	15 ± 1.8	2.3 ± 0.569

- Crystallite size increases with holding time – MBN energy decreases
- Carbides distributed throughout the matrix (EDS)
- With increase in holding time, carbides segregate near the boundaries and increase in size – soft matrix

X-Ray Diffraction

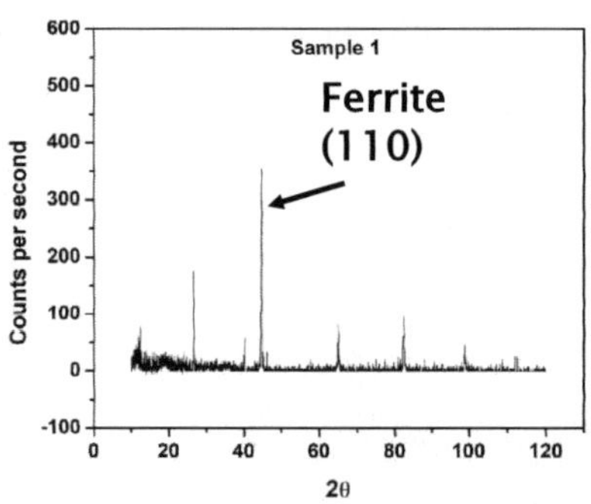

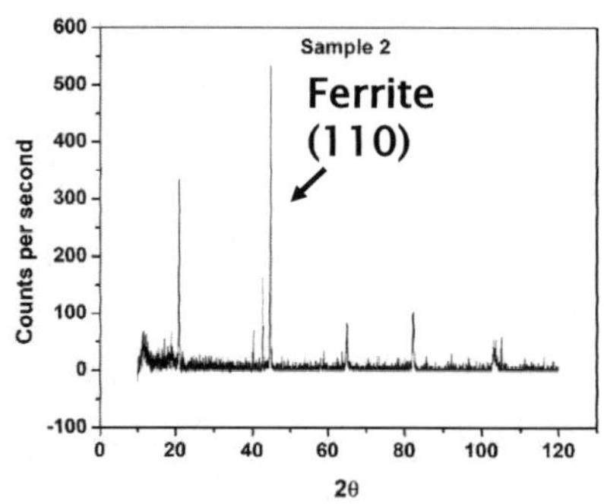

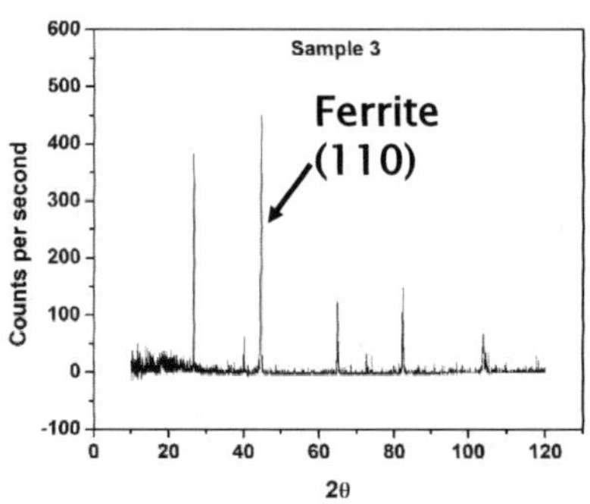

Ferrite peak (110) 2θ = 44.5	Sample 1	Sample 2	Sample 3
FWHM	0.20	0.15	0.16

Concept map

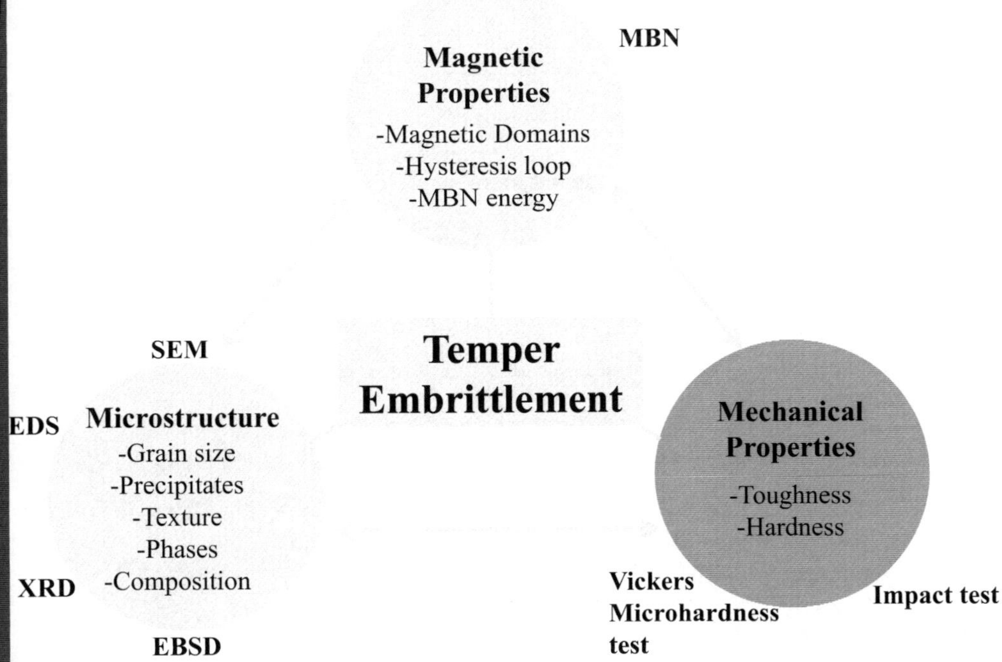

NDT in Canada 2018 | June 19-21 | Halifax, NS

23

Impact energy

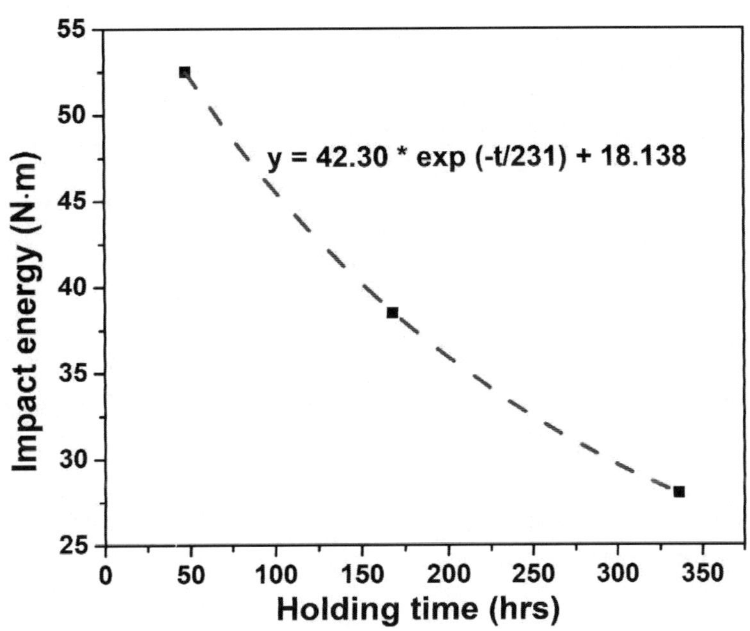

NDT in Canada 2018 | June 19-21 | Halifax, NS

24

Impact energy vs MBN energy

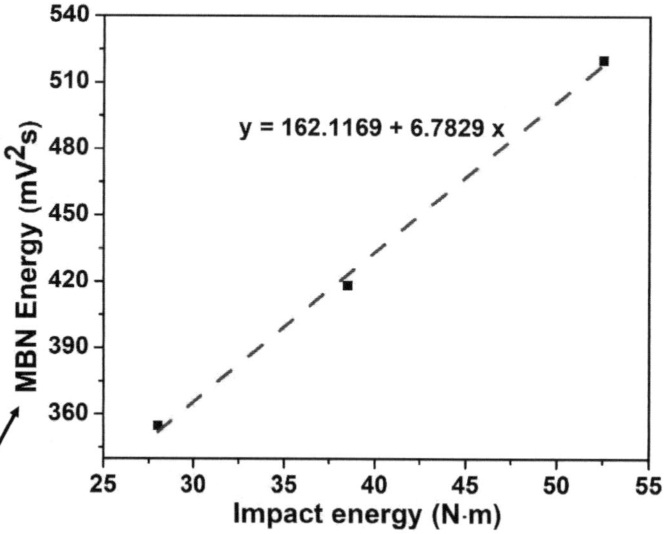

Average MBN energy for Surface 1 and Surface 2

Hardness - cross section

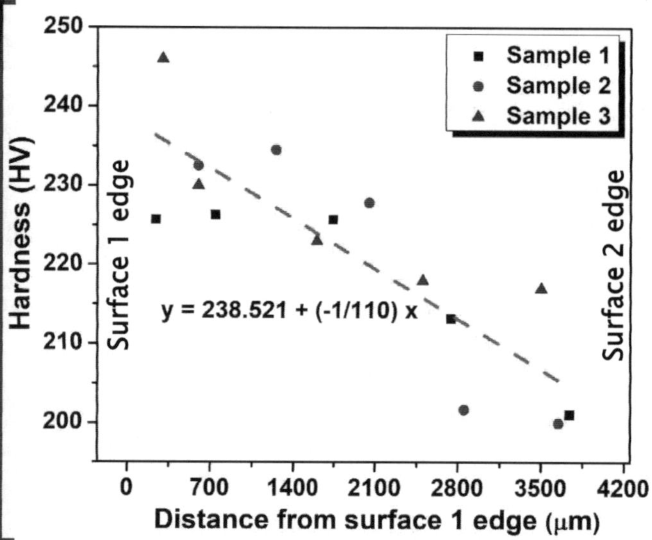

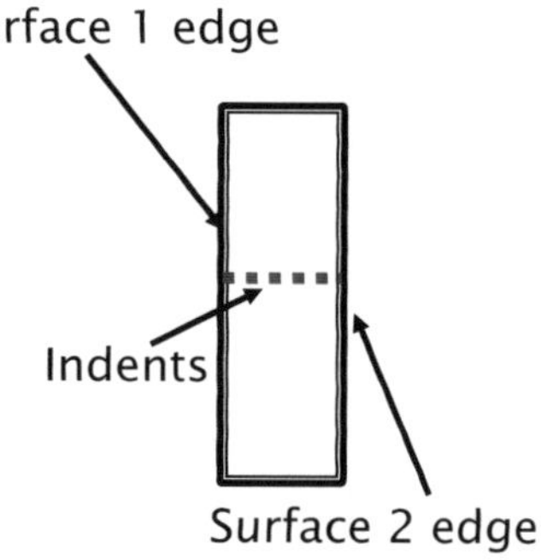

Sample cross-section

Hardness vs MBN energy

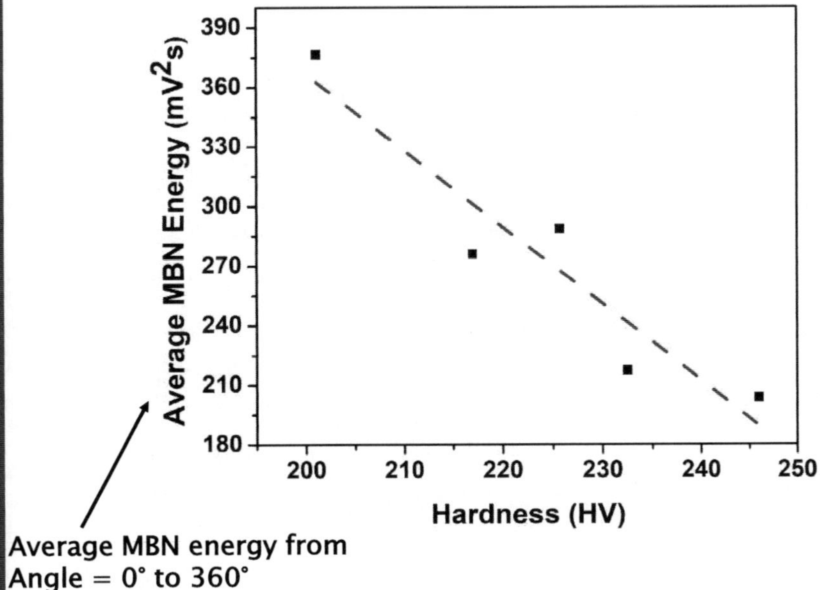

Hardness - cross section

- Carbide segregation near Surface 1 results in higher hardness near Surface 1 than Surface 2
- Carbides uniformly distributed - pinning - MBN energy isotropic for Surface 1
- Texture effect on MBN more dominant than carbide effect - Surface 2

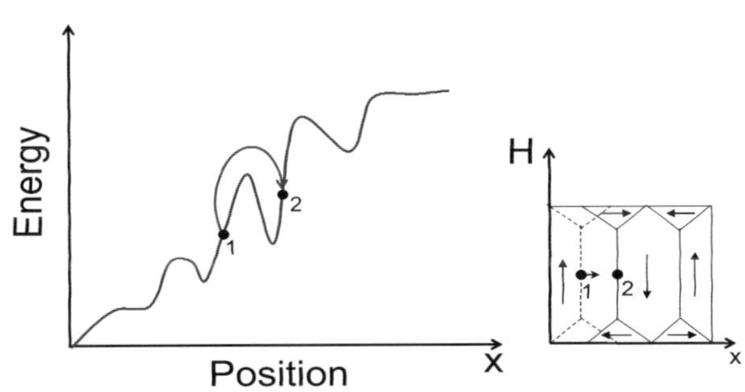

Krause et al. , Micromagnetic Techniques, ASM Handbook, 2018

Summary

MBN signal decreases exponentially

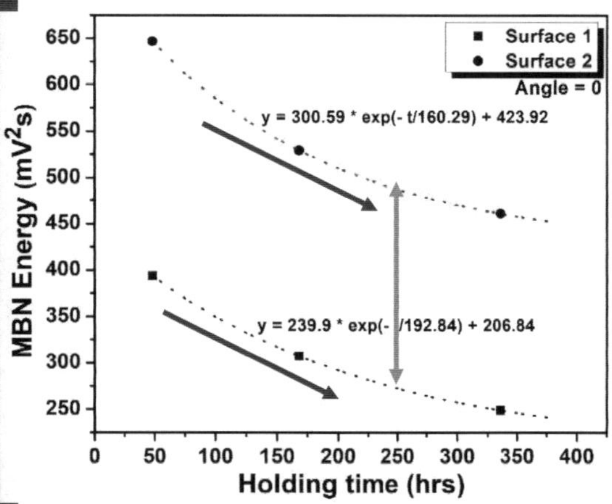

- Increase in Crystallite size
- Carbides – near the boundaries – soft matrix
- Migration of trace elements towards grain boundaries (P, Sn, As)

MBN energy ↔ Toughness

- Carbide segregation near surface 1 – pinning effect
- Hardness higher for Surface 1
- Texture effects – Surface 2

Acknowledgements

NSERC CRSNG

CDARP
(Canadian Defence Academy Research Program)

Thank you

NDT in Canada 2018 | June 19-21 | Halifax, NS

Digital Radigraphy in Aerospace

Underlying principles and migration from film technology

CINDE
CANADIAN INSTITUTE FOR NON-DESTRUCTIVE EVALUATION

NDT in Canada
NDTiC 2018
Canada's NDT Conference

June 19 - 21
Halifax Convention Centre
Halifax, Nova Scotia

- The big challenge?

NDT in Canada 2018 | June 19-21 | Halifax, NS

Different applications

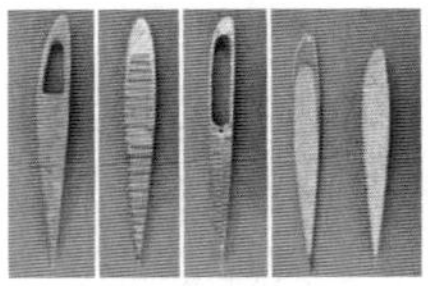

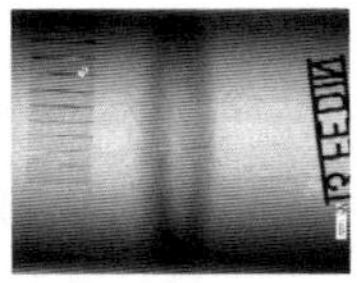

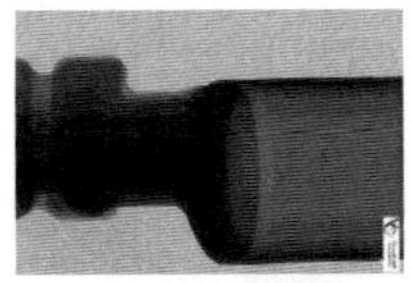

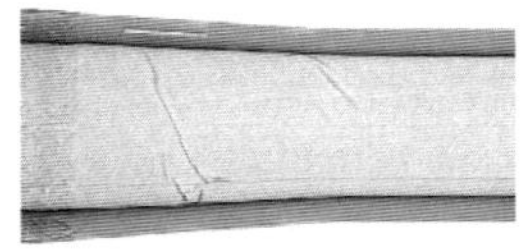

- Disbonds
- Cracks
- Delaminations

- Cracks
- Pores
- Inclusions

- Porosities
- Inclusions
- Geometry

- Delaminations
- Pores
- Inclusions

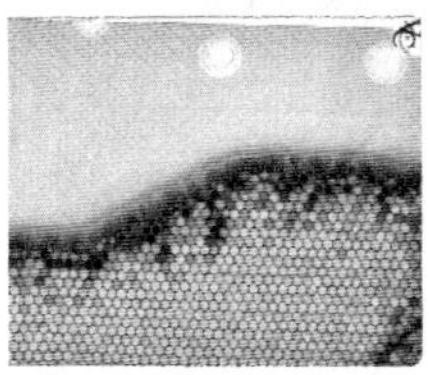

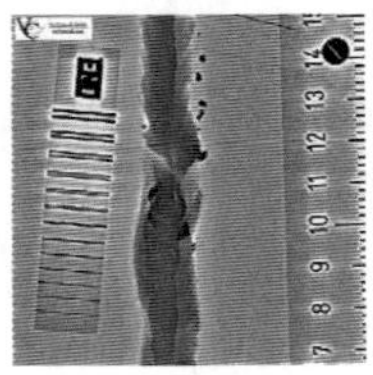

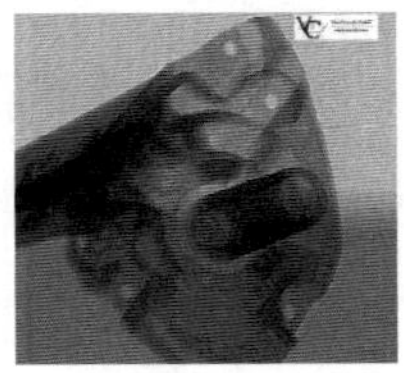

NDT in Canada 2018 | June 19-21 | Halifax, NS

X-ray tube types

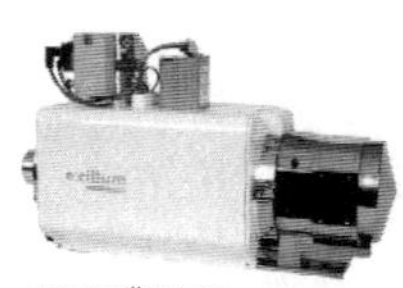

	Nanofocus-Tube	Microfocus-Tube	Minifocus-Tube
Focal spot size d	150 nm	2 µm - 300 µm	0,3 - 1 mm
Voltage U	Up to 190 kV	Up to 450 kV	Up to 600 kV

NDT in Canada 2018 | June 19-21 | Halifax, NS

Focal spots

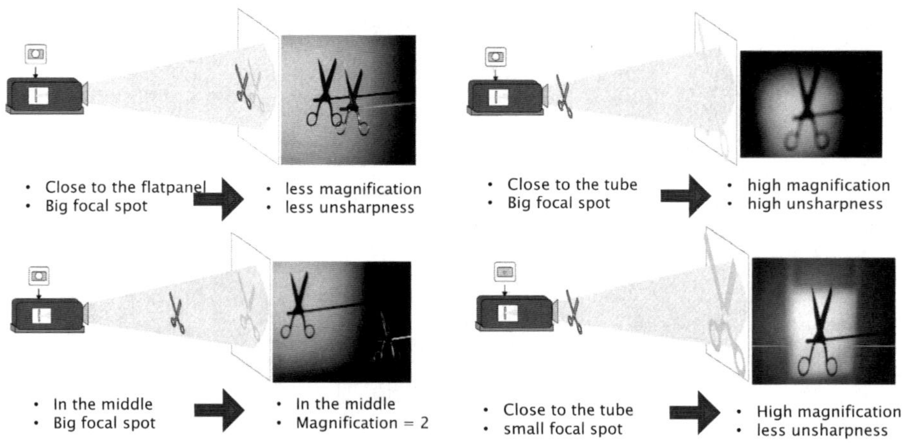

NDT in Canada 2018 | June 19-21 | Halifax, NS

Basics of X-ray inspection

FLAT PANEL

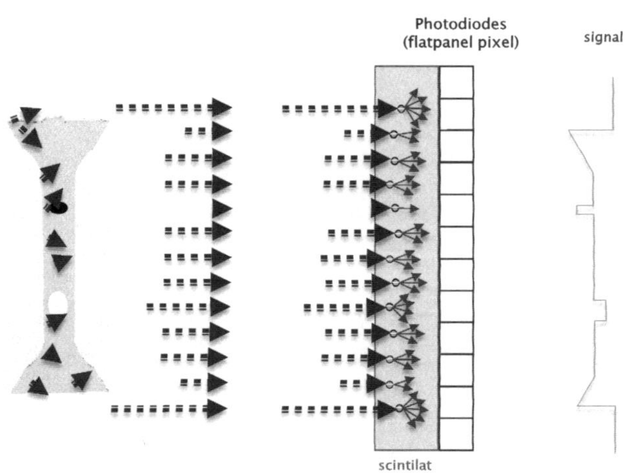

NDT in Canada 2018 | June 19-21 | Halifax, NS

Standards and qualification

- Customer requirements
  - Boeing (BSS 7044), Airbus (AITM), etc.
- Industry requirements
  - NADCAP, ISO 9100 etc.
- Method standards
  - ISO, ASTM, etc.

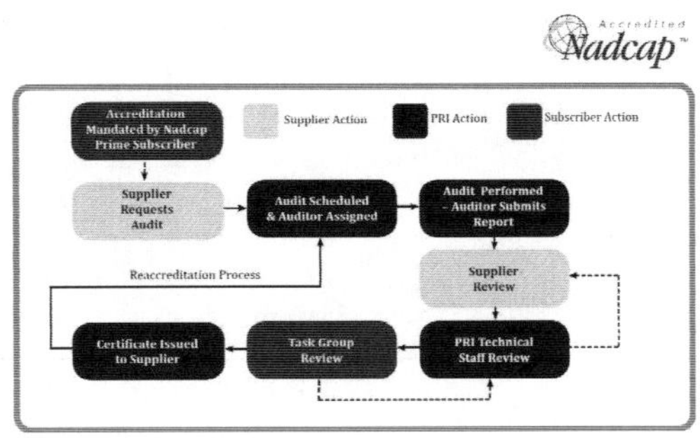

System Performance Checks

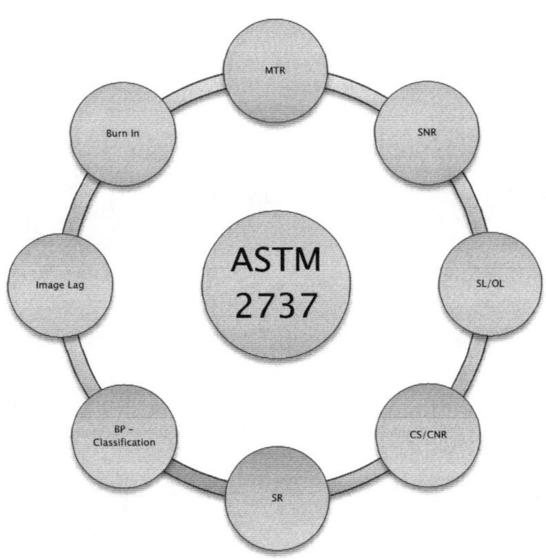

Long-term eval. - ASTM2737

- DR phantom for detector evaluation
- Containing: Step wedge, two hole penetrometers, 2 Duplex IOI's
- Standardized performance evaluation
- Report generation according to ASTM
- Other company specific phantoms available as well

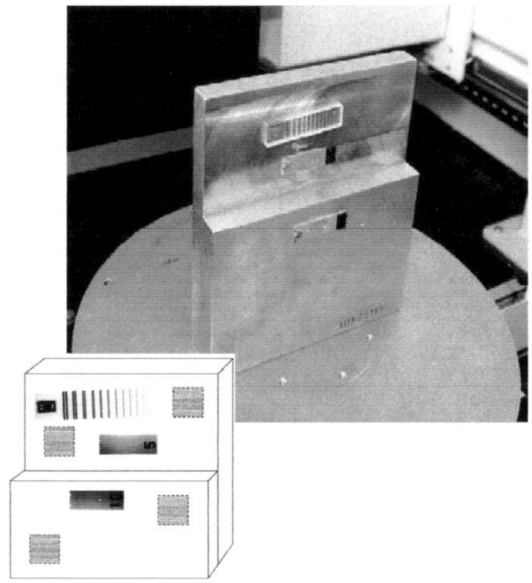

Spatial Resolution (SRb)

<u>IMAGE QUALITY - RESOLUTION</u>

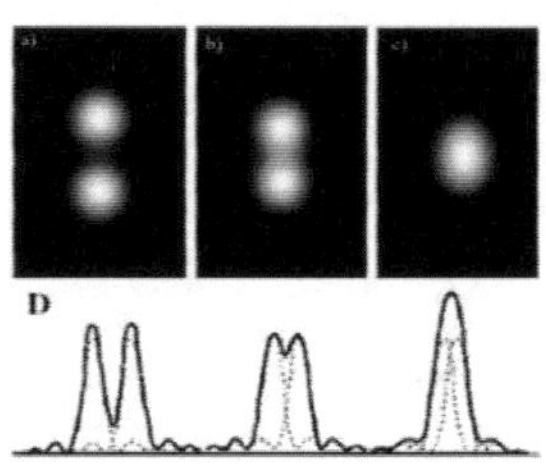

http://physwiki.apps01.yorku.ca/images/thumb/3/3e/Fig15_airy.png/400px-Fig15_airy.png

https://developer.apple.com/ios/human-interface-guidelines/icons-and-images/image-size-and-resolution/

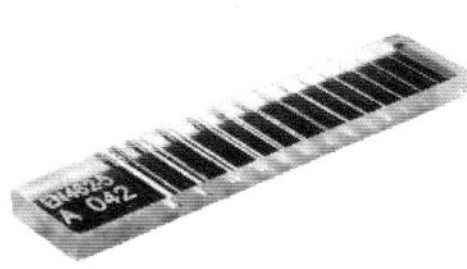

Duplex Image quality indicator (IQI)

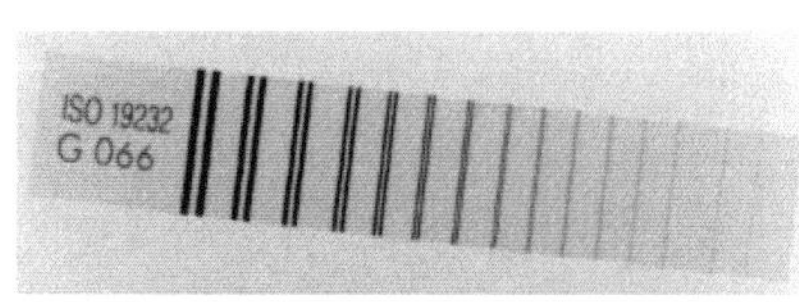

Spatial Resolution (SRb)

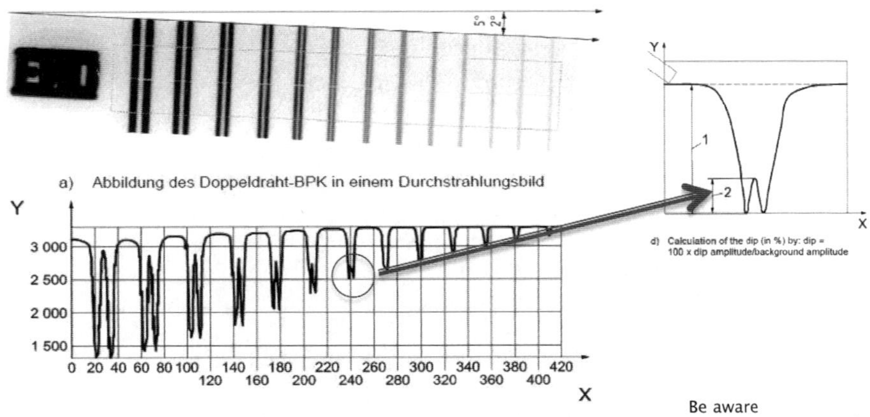

b) Aus mindestens 21 Zeilen gemitteltes Profil des Doppeldraht-BPK

Be aware
The dip (2) must be a minimum of 20% of the maximum (1)

This calculation has to be done by Software and not by human eyes

Spatial Resolution (SRb)

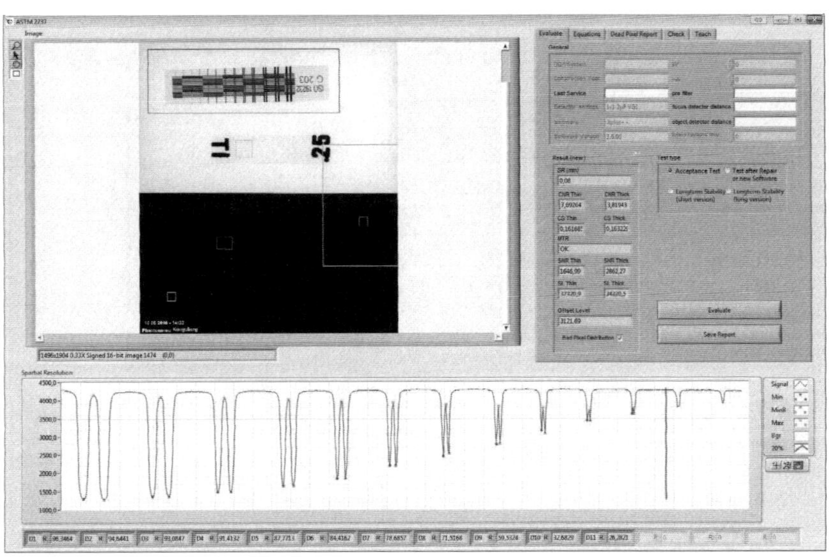

Contrast Sensitivity (CS)

Contrast Sensitivity (CS)

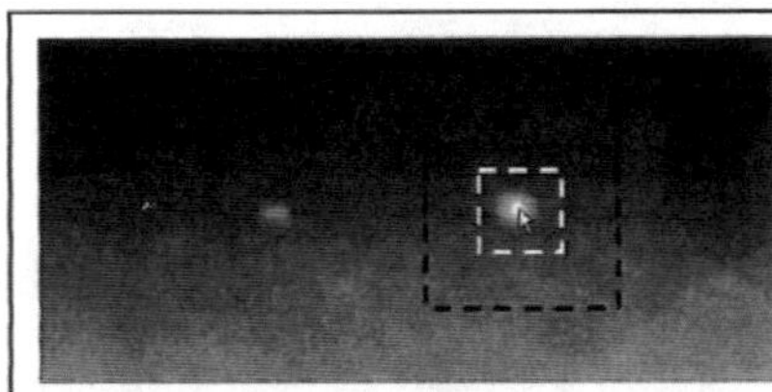	$CNR = \frac{GV_{median}[hole] - GV_{mean}[beside\ squares]}{Sigma[beside\ squares]}$ $CS\ [\%] = \frac{GBV}{CNR} * \frac{MT_{IQI}}{MT_{total}} * 100$ Known Parameter: $MT_{step} = 50mm$; $MT_{IQI} = 0.5mm$ GBV = 2.5
Measured Parameter: GV_{median} [hole] = 4800 GV_{mean} [beside squares] = 4600 Sigma [beside squares] = 50	Calculation: $MT_{total} = 50.5mm$ CNR = 4800 - 4600 / 50 = 4 CS [%] = 2.5 / 4 * 0.5mm/50.5mm * 100 CS [%] = 0.6188%

Signal to Noise (SNR)

Tabelle D.1 — Erforderliche $SNR_{measured}$-Werte für ausgewählte CR-Systeme mit unterschiedlichen SR_b, die den jeweiligen SNR_N-Werten äquivalent sind

System-parameter	Hochauflösendes System			Standardsystem					
Doppeldraht-BPK-Qualifizierung	13+	13	12	11	10	9	8	7	6
Basis-Ortsauflösung SR_b	40 µm	50 µm	63 µm	80 µm	100 µm	130 µm	160 µm	200 µm	250 µm

$$SNR_N = SNR_{measured}\,\frac{88{,}6\mu m}{SR_b}$$

$$\text{SNR-measured} = \frac{\text{Meanvalue}}{\text{Standard deviation}}$$

From the image

Report generation

DDA System		voltage	kV	
Construction Year		tube current	mA	
Last Service		pre filter (material and thickness)		
Detector Settings		focus detector distance	mm	
Software		object detector distance	mm	
Software Version		total exposure time per image	s	

Test	☐ Acceptance Test	Used IQIs	☐ 5-Groove-Wedge	Material of the used IQIs	☐ Aluminium
	☐ Test after Repair or new Software		☐ Duplex Plate Phantom		☐ Titanium
	☐ Longterm Stability (short version)		☐ Duplex Wire IQI (EN 462-5)		☐ CRES
	☐ Longterm Stability (long version)		☐ Hole IQI		
			☐ Wire IQI (EN 462-1)		
			☐ no IQI required		

Tests		Unit	Result (new)		Limit		Result		Remark
			thin	thick	thin	thick	thin	thick	
Spatial Resolution	SR	µm							
Contrast Sensitivity	CS	%							
Material Thickness Range	MTR	mm							
Signal-to-Noise Ratio	SNR								
Signal Level	SL								
Image Lag	Lag	%							
Burn In	BI	%							
Offset Level	OL								
Bad Pixel Distribution									

Date of Tests	
Conclusion	
Operator	

Bad pixel report – ASTM 2597

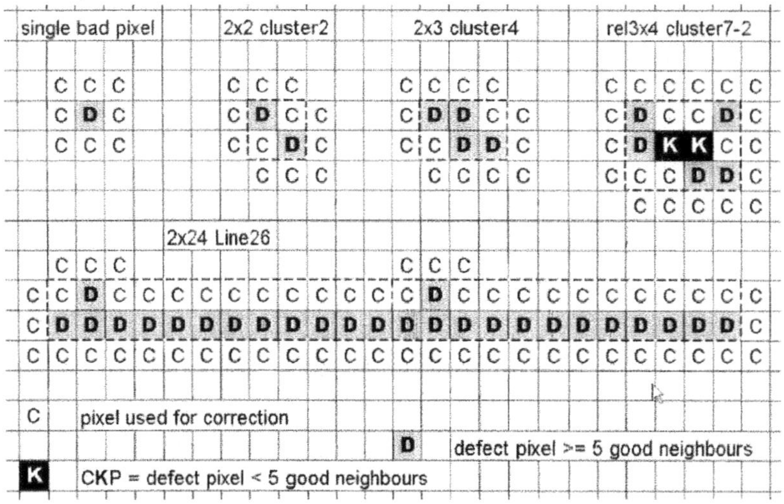

Bad pixel report – ASTM 2597

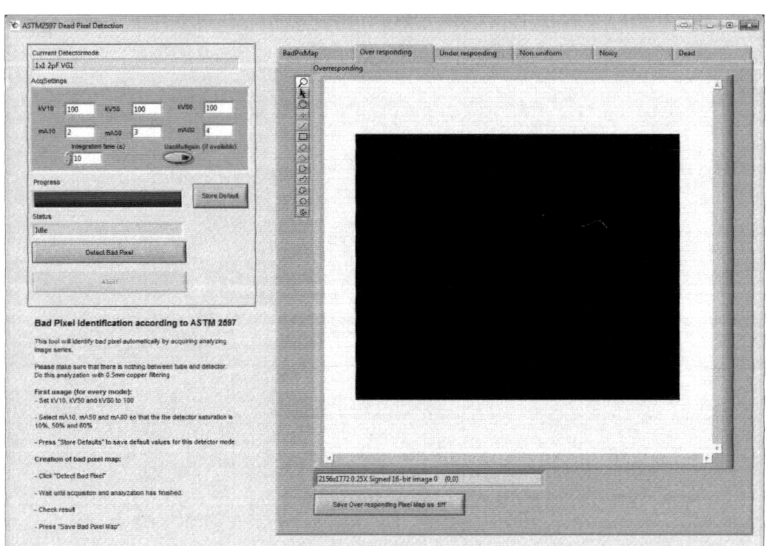

Alternative: TAM Phantom

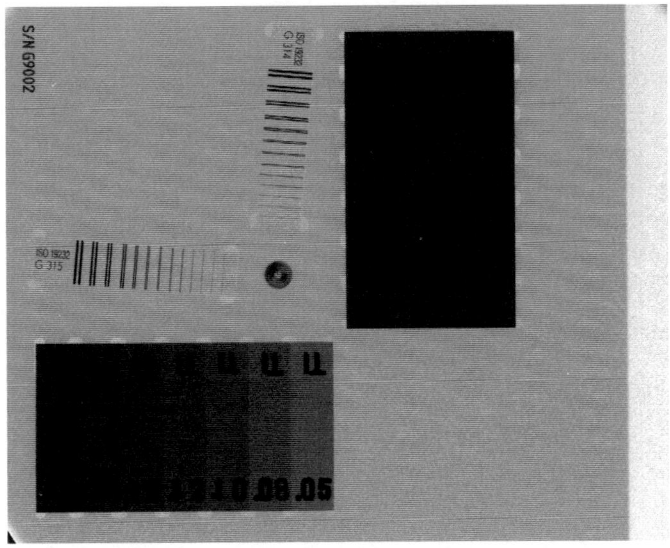

Digital filters

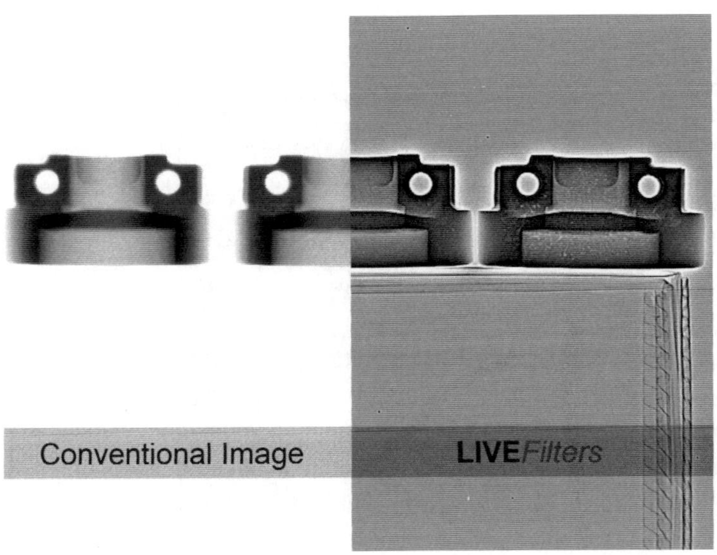

NDT in Canada 2018 | June 19-21 | Halifax, NS

Measurement

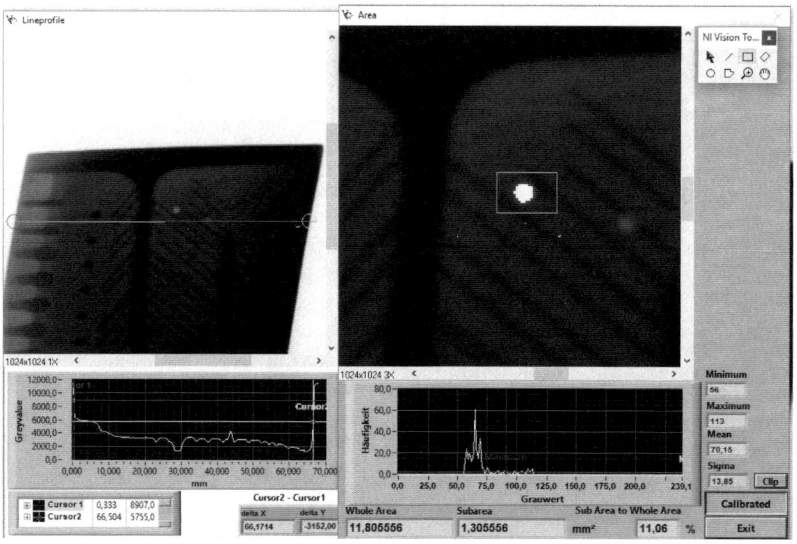

NDT in Canada 2018 | June 19-21 | Halifax, NS

Digital references images

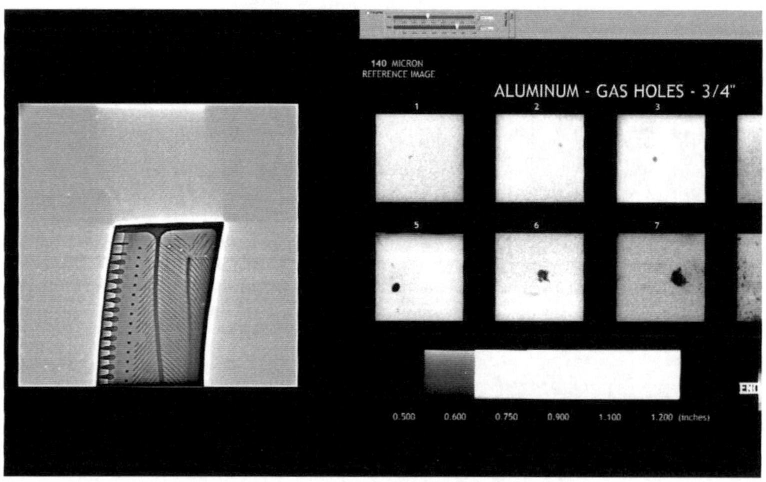

NDT in Canada 2018 | June 19-21 | Halifax, NS

Digital reference images

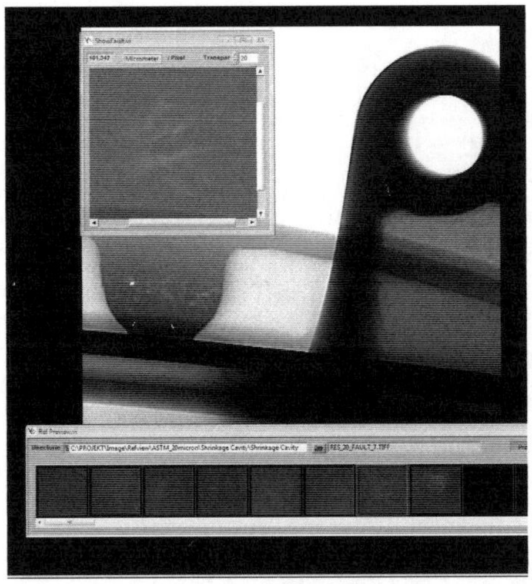

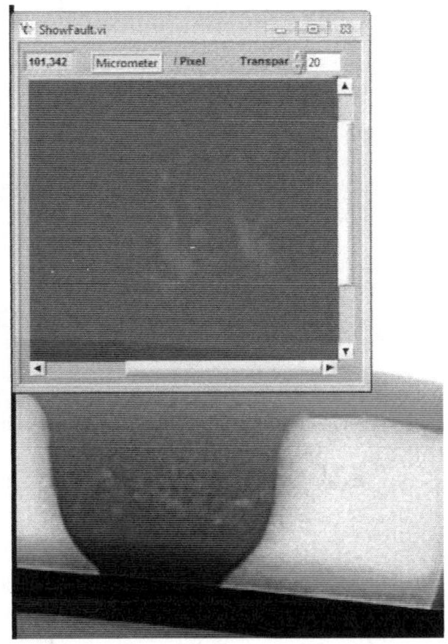

NDT in Canada 2018 | June 19-21 | Halifax, NS

Digital overlays

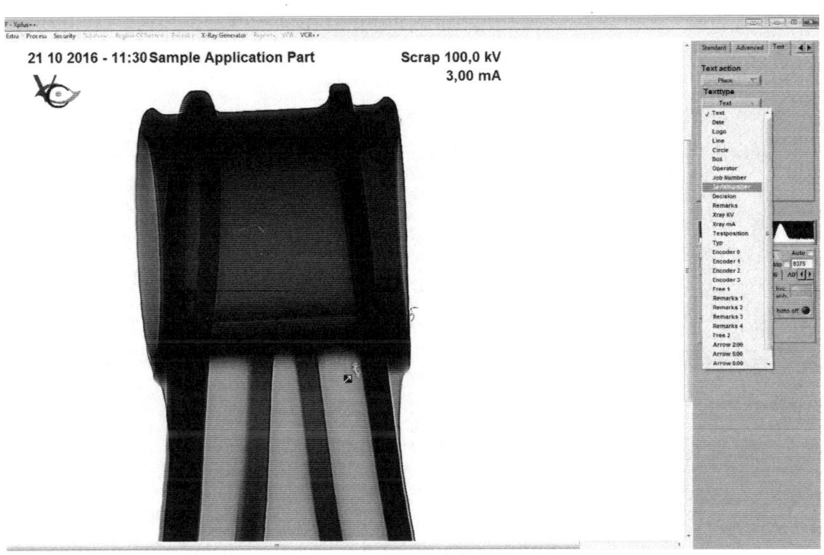

NDT in Canada 2018 | June 19-21 | Halifax, NS

Computed Tomography

- QUICKScan

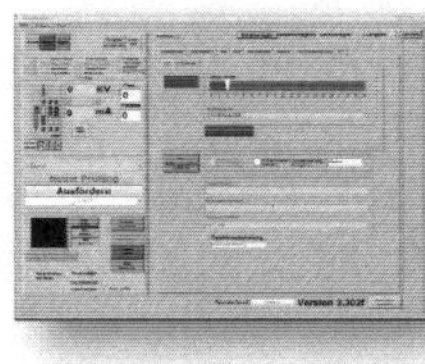

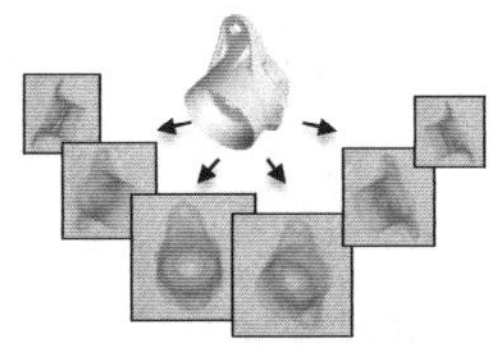

- Reconstruction and analysis

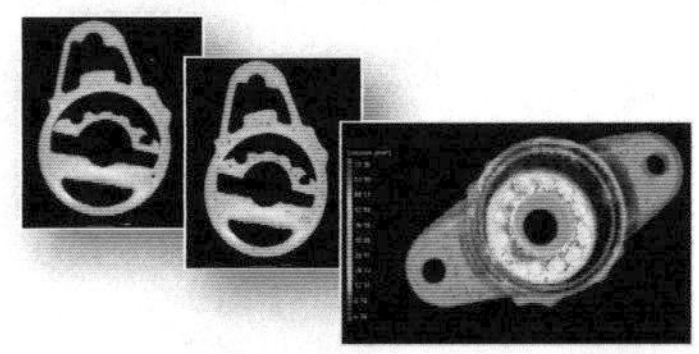

- Statistics and decisions

Material	
Materialvolumen [mm³]	100654.29
Defektvolumen [mm³]	1160.60
Verhältnis Defekte/Gesamtvolumen [%]	1.14

Compute Tomography

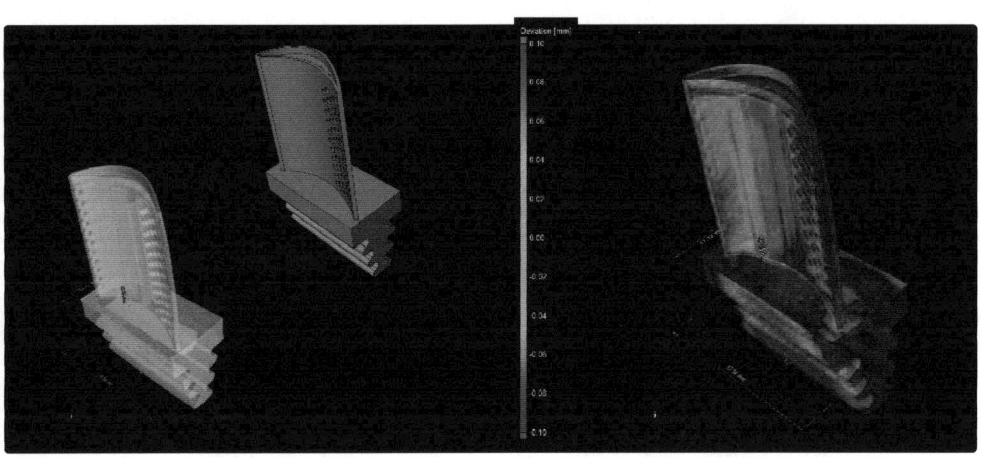

Commercial justification

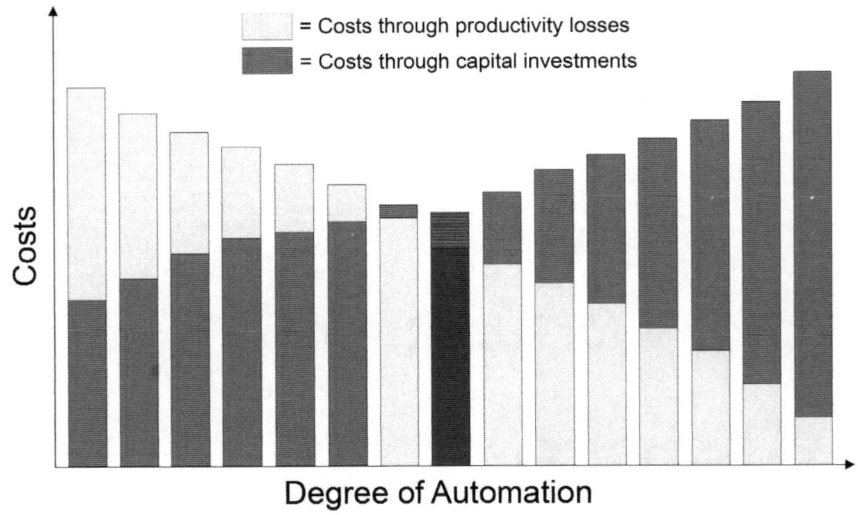

NDT in Canada 2018 | June 19-21 | Halifax, NS

Commercial justification

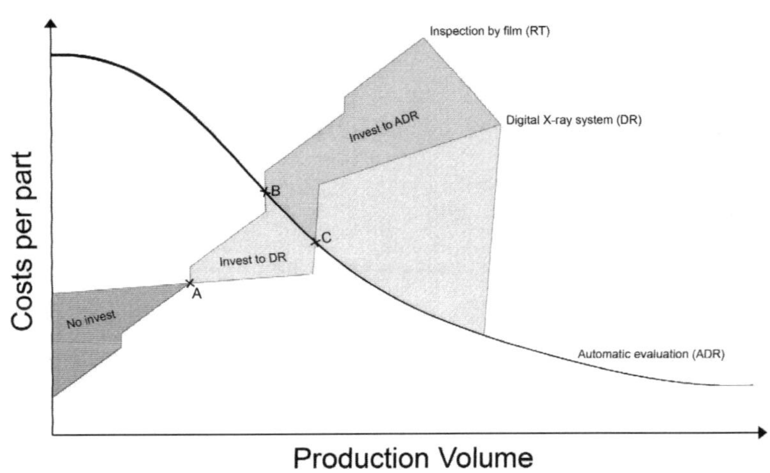

NDT in Canada 2018 | June 19-21 | Halifax, NS

Thanks for your attention

VisiConsult

X-ray Systems & Solutions GmbH

+49 (0) 451 2902860

l.schulenburg@visiconsult.de

NDT in Canada 2018 | June 19-21 | Halifax, NS

Accelerator Based Large Format Computed Tomography

CINDE
CANADIAN INSTITUTE FOR NON-DESTRUCTIVE EVALUATION

NDT in Canada
NDTiC 2018
Canada's NDT Conference
June 19 - 21
Halifax Convention Centre
Halifax, Nova Scotia

TABLE OF CONTENTS

Who is Jesse Garant Metrology Center
Why Use Computed Tomography
Construction of our additional Facility
Custom Large Format DDA Panel
Linac Testing
Real World Results

NDT in Canada 2018 | June 19-21 | Halifax, NS

WHO IS JESSE GARANT METROLOGY CENTER

JESSE GARANT
METROLOGY CENTER

Jesse Garant Metrology Center provides solutions that helps create meaningful advancements in industry. By focusing on advanced imaging solutions such as Industrial Computed Tomography with the support of production x-ray and 3d scanning services, we deliver critical insights into inspection projects that are difficult to measure. With over 15,000 unique parts CT scanned, spanning eight years in over ten different industries, we continue to invest in our Industrial CT Scanning Services and have the proper infrastructure in place to operate efficiently.

WHY USE COMPUTED TOMOGRAPHY

1. Cabinet/Vault

A shielded enclosure that houses all the components of an xray/CT which also prevents minimum leakage of radiation to the surrounding area.

Can vary in size and weight depending on the amount of radiation produced by the source

2. X-Ray Source

X-ray sources come in a variety of energy settings with different spot sizes. The key points for an x-ray source are:

Source power and voltage
Spot size
Consistent power setting

3. Precision Rotary Table Positioner

The rotary table is required to capture the projects over a 360 degree arc.

This distance needs to be known with high precision in order to get accurate dimensions

4. Detectors

A digital detector is a device that captures x-rays, converts them to light and then recorded. The inverse of what computer monitors do.

Digital detectors come in a variety of pixel pitch's which mean the space between pixels.

NDT in Canada 2018 | June 19-21 | Halifax, NS

VAULT CONSTRUCTION PHASE

- The base vault construction took just over 4 months to assemble and complete.
- Walls are over 6' (1828.8mm) thick
- The weight of the vault topped out at 1.8 million lbs.
- The door alone weighs 14000 lbs and requires a 3HP motor to move it.

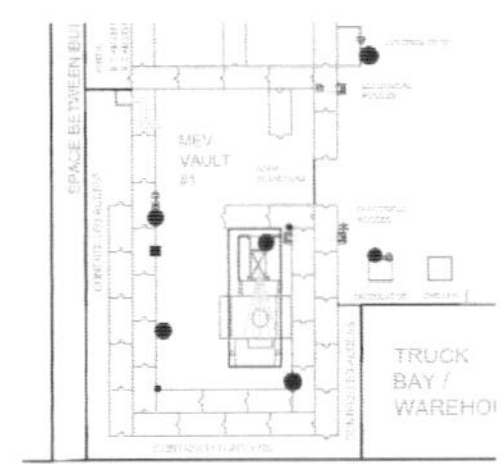

NDT in Canada 2018 | June 19-21 | Halifax, NS

VARIAN M3A LINAC

- Modulator Cabinet
  - Converts and monitors the main electrical power to high voltage pulses to the magnetron
- Magnetron
  - Provides high frequency microwave power used to accelerate the electrons in the waveguide.
- Waveguide
  - Tube that guides the electromagnetic waves from the magnetron to the accelerating guide where the electrons become accelerated.
- Target
  - Usually made from tungsten for industrial purposes is located at the end of the accelerating electrode, used to convert the electrons to Xrays
- Chiller/Water Cooling System
  - Required for the operation of the system and allows the components to operate constant temperature

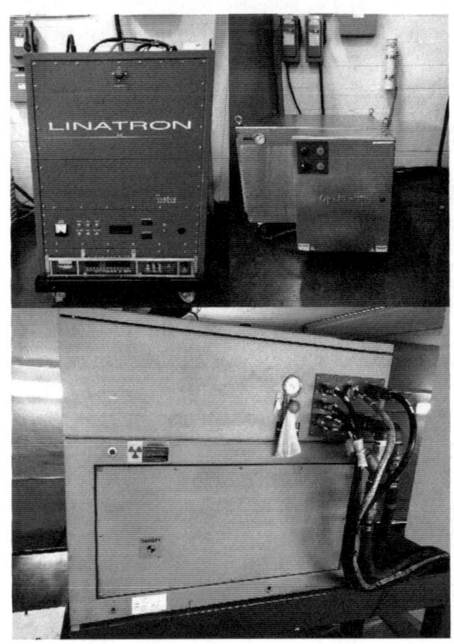

NDT in Canada 2018 | June 19-21 | Halifax, NS

DIGITAL DETECTORS

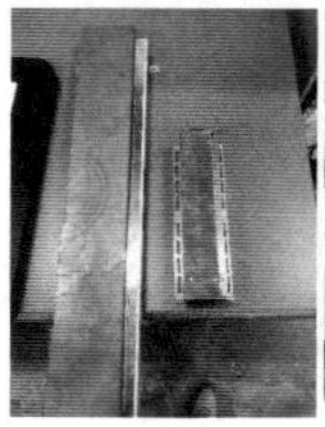 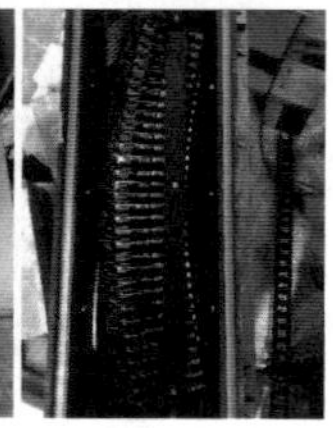

 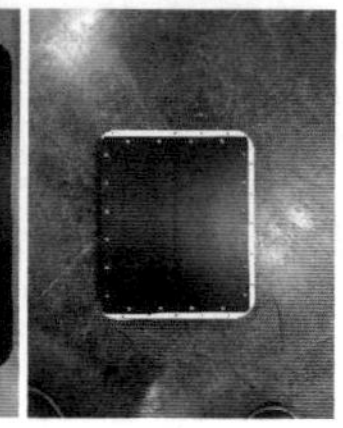

Linear Detector Arrays (LDAs)

LDAs are one-dimensional x-ray detectors that consist of a single row of x-ray sensitive detectors. In order to obtain a two-dimensional image from such a detector it is necessary to move the detector, or the object in a direction perpendicular to the length of the array. In order to maintain regular image geometry (square pixels), the speed of movement has to be carefully synchronized with the line exposure time of the array. The width of two-dimensional images acquired by an LDA are limited by the number of pixels comprising the width of the array, however, the length is unlimited. LDAs are also suited to perform 2D CT which obtain single CT slices through an object. Although slower, 2D CT is preferable to 3D CT (using an area detector) for applications where scattered x-rays are a problem; for example for high energy applications, radioactive components or when examining dense materials.

Digital Detector Arrays (DDAs)

DDA's are composed of two principle imaging components. The first imaging element is a scintillator screen that converts the x-rays into visible light. A wide range of scintillation screens and materials are suitable and most manufacturers offer different screen options. This screen is coupled to an array of photosensitive photodiodes which convert the visible light to an electrical charge. This charge can be read out via an array of switches of Thin Film Transistors (TFT's) and digitized. The digitized values for each photodiode, can be used to generate a corresponding grey level value in single image pixel. Thus the array of photodiodes allow a digital radiograph to be captured, each diode corresponding to a pixel in the radiographic image. However currently there is manufacturing limitations in regards to size. The largest most common panels are approximately 16" in size.

CUSTOM BUILT PANEL

- The panel is a custom built 1 meter by 1 meter panel.
- Has an effective resolution capable of doing 2048x2048 @ 16bit (500um).
- Capable of withstanding energies up to 6MeV or 8 Gy/min
- Mountable to either an actuator or simply on feet.
- Fast integration time of 10ms

CUSTOM BUILT PANEL

- Due to the size and the construction of the detector it is subject to Negative Optical Distortion (barrel distortion).
- The barrel distortion caused from the distance from the point source to the detector from an un flattened beam needed to be corrected.
- We took a precision hole patter which we first binarized and used a special algorithm to correct for the optical distortion.

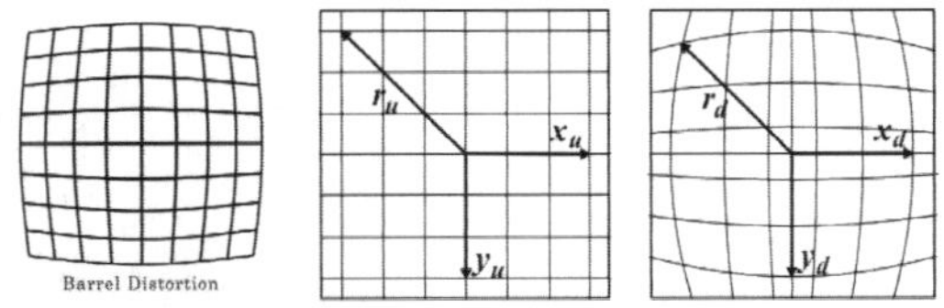

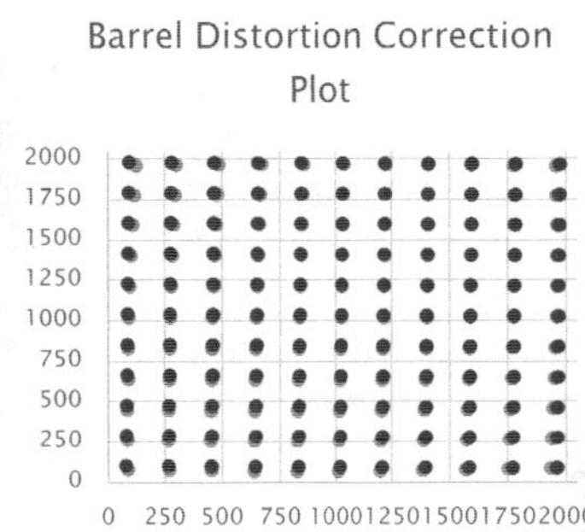

HOW DOES IT ALL WORK TOGETHER?

The process of computed tomography is that it takes several thousands 2D x-ray images called projections covering a total of 360 degrees.

The object is fixtured in a low density material to prevent movement during the 360 degree rotation.

These projections are required for the reconstructor to create a 3d dimensional dataset (VOXEL).

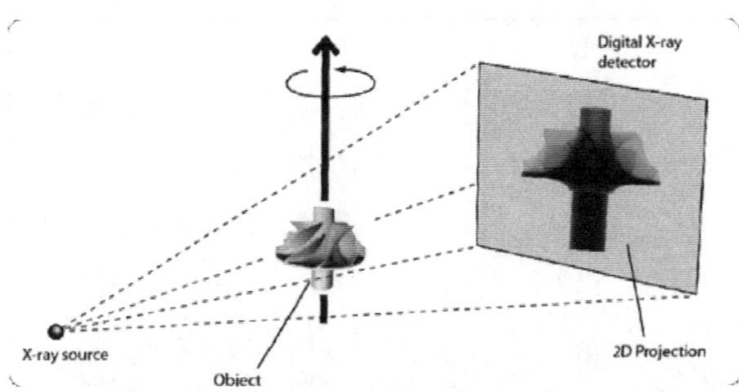

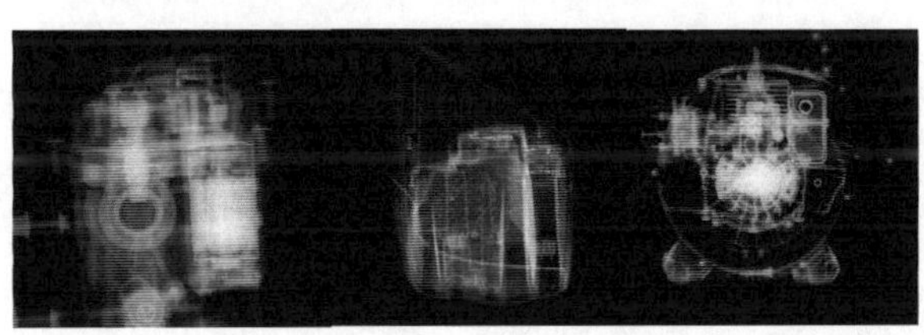

USING VOLUME GRAPHICS TO RECONSTRUCT ALL DATA

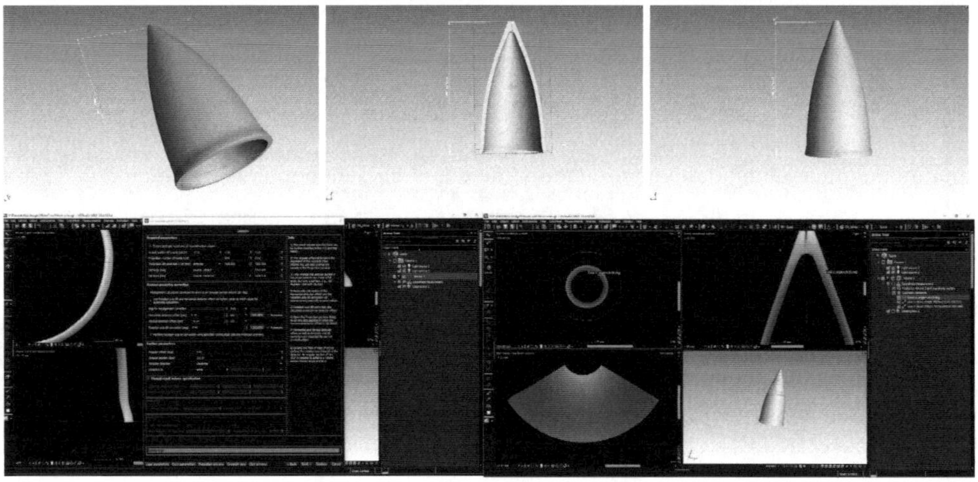

NDT in Canada 2018 | June 19-21 | Halifax, NS

LINAC TESTING

LINAC TESTING PHASE

- With the reconstruction tools in VG this allowed us to apply different scanning techniques (Field of View Enlargement, HELIX, etc)
- Our scans using the large format panel yielded little to no scatter from materials of different densities.
- Connections and welds can be seen quite clearly between dissimilar materials and as accurate as 500 microns

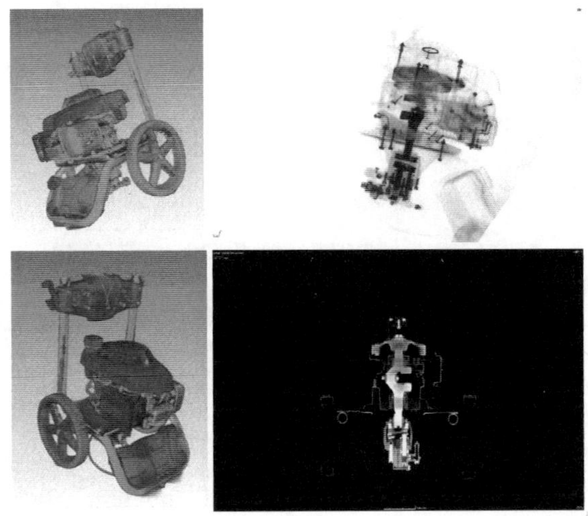

TESTING PHASE

NDT in Canada 2018 | June 19-21 | Halifax, NS

TESTING PHASE

Ever wonder how full your ShopVac is without opening it?

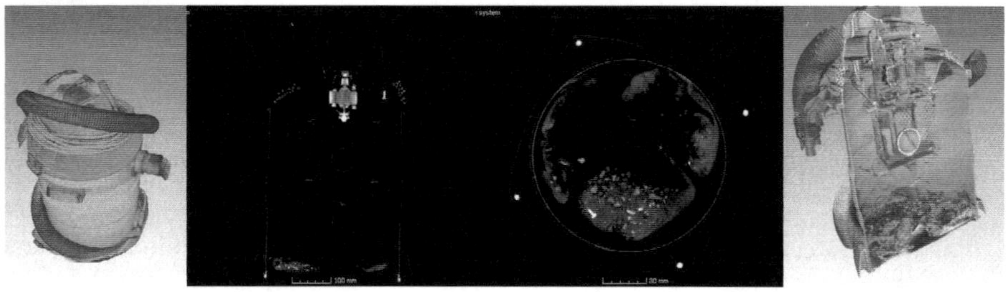

SO WHY USE A LINAC?

Benefits of using a Linear

Accelerator for CT:

- The ability to see clean interfaces between low density and high density material.
- The grayscale of the low energy scan; the material doesn't correspond to the real material as shown in the MeV scan data.

A UPS Power Supply Scanned on 3MeV (Left) & 450KeV (Right)

SO WHY USE A LINAC?

Benefits of using a Linear

Accelerator for CT:

- High energy radiation is less likely to be scattered due to the energy removed by the scattering of the photon. In short basically means its more likely to continue to travel in the direction intended.
- As the energy of photons increase the probability of interaction drops.
- The imaging quality great increases due to the high signal homogeneity and lower beam hardening artifacts.

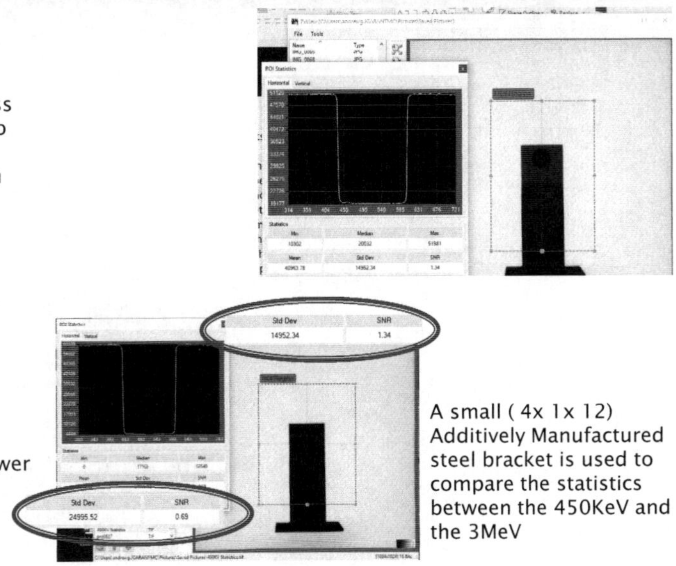

A small (4x 1x 12) Additively Manufactured steel bracket is used to compare the statistics between the 450KeV and the 3MeV

SO WHY USE A LINAC WITH DDA? TIME!

- Going to a large format flat panel has drastically improved the acquisition time at little to no reduction in image quality for most parts.
- If using a 2048 pixel horizontal line, Nyquist states that we should be using 3201 frames for proper acquisition.
- Typical LDA run time at 3201 projections per slice @ 90 seconds per slice at .5mm increments for a total of 915 slices for the casting. This in turn equates to **22 hours!**
- With the large format DDA panel we are down to 27 minutes for 3201 frames for the same 2K matrix.
- There is little to no discernable loss of quality and no more missed defects on the Z axis

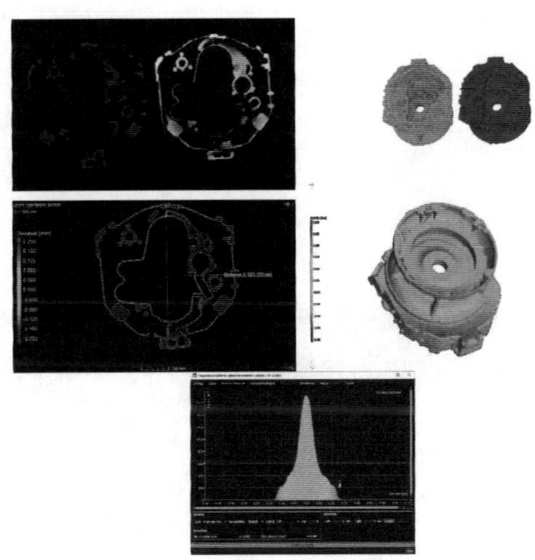

REAL WORLD EXAMPLES

LARGE FORMAT METAL PARTS

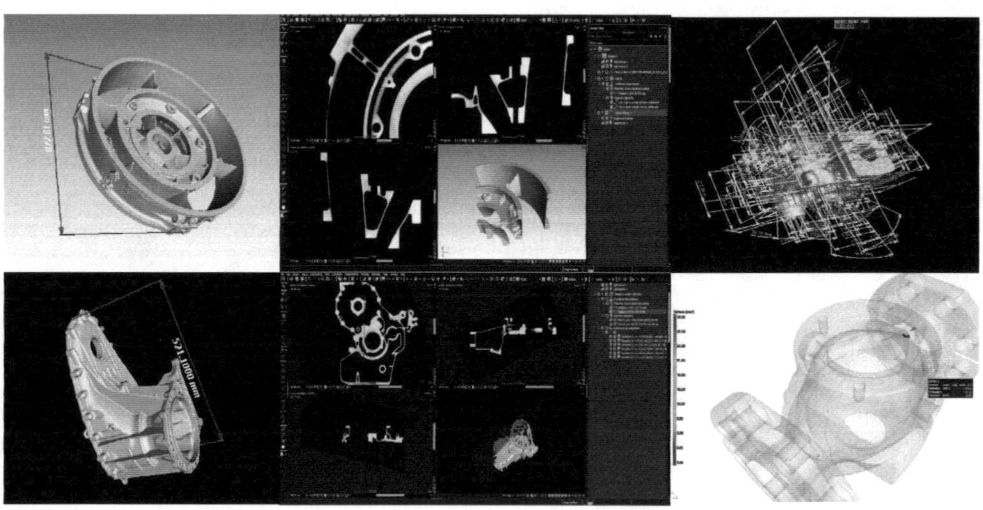

LARGE FORMAT METAL PARTS - MEASURMENTS

Once a part has been scanned, a Part to CAD comparison is first performed and aligned by the part print datums. The developed measurement plan is then executed. Results for every programmed GD&T point are provided in an easy to read chart or in excel format. The first article inspection results show the listed dimension, tolerance, deviation from actual, and value out of tolerance. Within the program, the customer is also able to go to any dimensioned feature and see how the dimension was taken.

•Perfect fit for PPAP or AS9102 form 3 reporting requirements
•Once a measurement plan has been developed and the part has been scanned, tens, hundreds, or thousands of part print points can be dimensioned instantly.
•Complex GD&T callouts such as profile, run out, MMC, LMC, and concentricity can easily be computed
•Complex parts no longer have to be cut up to perform full part print layouts
•Multiple parts can be dimensionally validated faster than traditional inspection methods for a first article PPAP part run

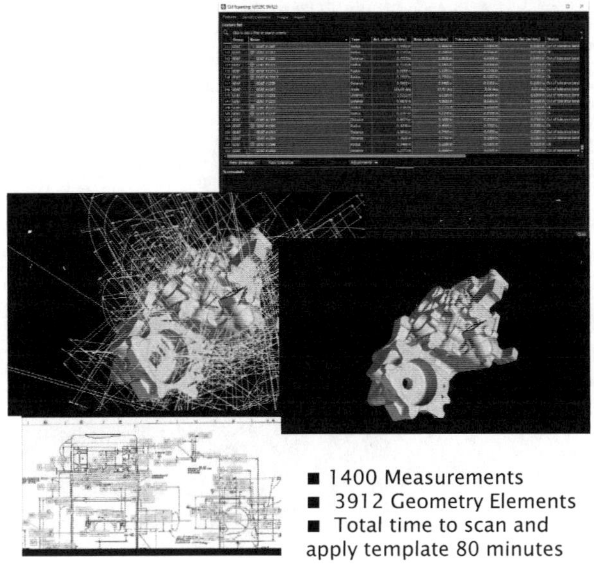

- 1400 Measurements
- 3912 Geometry Elements
- Total time to scan and apply template 80 minutes

**Template previously created

NDT in Canada 2018 | June 19-21 | Halifax, NS

MEASURMENTS – TEXTURES – SCRIBE LINES

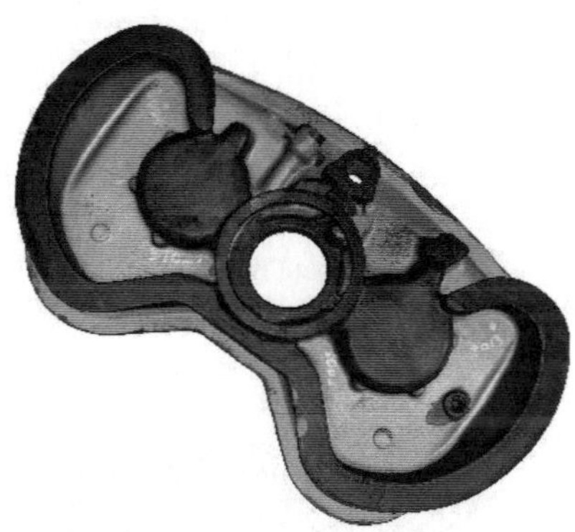

NDT in Canada 2018 | June 19-21 | Halifax, NS

LARGE FORMAT METAL PARTS – PART TO CAD/PART

After a part is CT scanned, a supplied CAD model or using another part from the customer the data is compared to the CT dataset. Alignment is performed by either a best fit method or predetermined datum's. Difference are then qualified by:

- The distance between surfaces
- Whisker Plots
- Tolerated Values

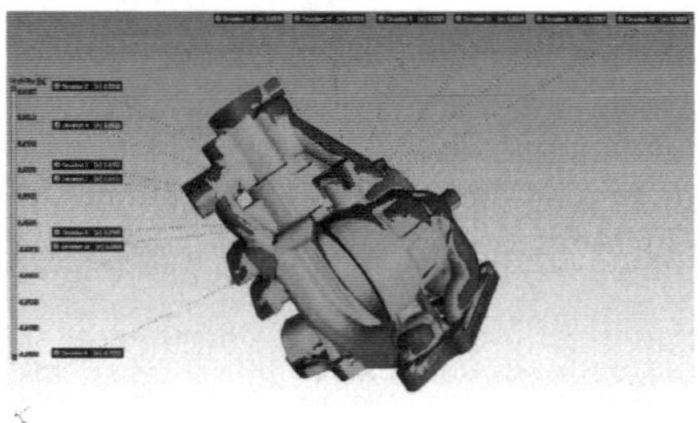

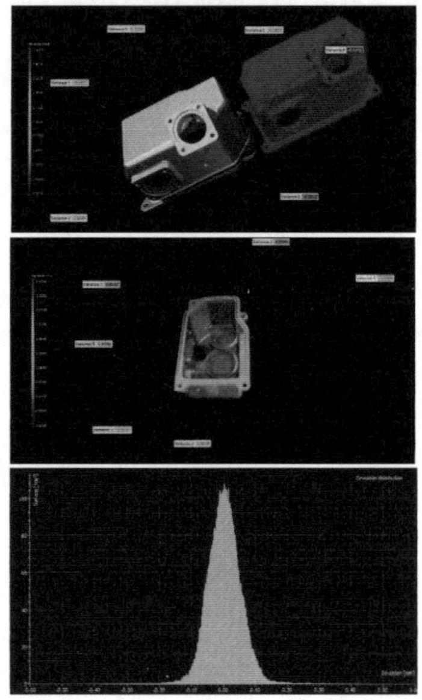

LARGE FORMAT METAL PARTS – WALL THICKNESS

After a part is CT scanned, a ISO surface boundary is created around all internal and external part features. The distance between every surface is calculated and identified as the minimum wall thickness. The results are supplied in a color coded model showing variations from a pre-specified wall thickness.

• Results can identify slight variations in a parts wall thickness, which can become costly to the manufacture in high volume part runs
• Results can identify thin walls not meeting the engineered wall thickness sizes for either structural or pressure requirements.

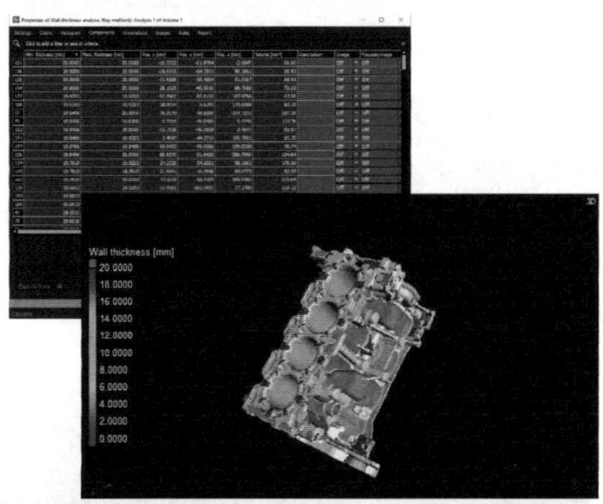

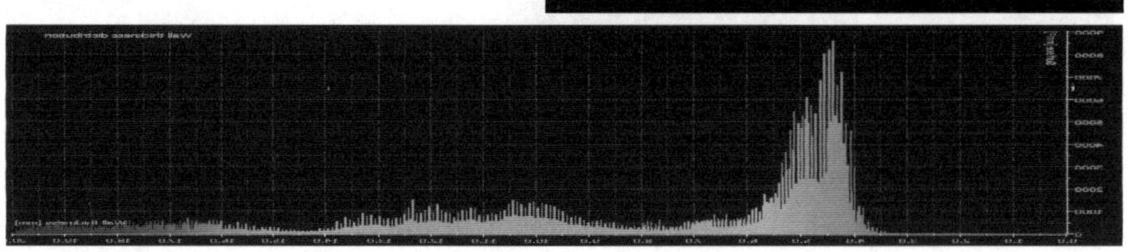

LARGE FORMAT METAL PARTS – VOID/INCLUSION

After a part is CT scanned, a ISO surface boundary is created around the part itself and for each internal porosity within the part. Internal porosity are then quantified by the volume of each pore. Once the volume of each pore has been identified, the results are assigned a color based upon volume and are indexed for easy location identification.

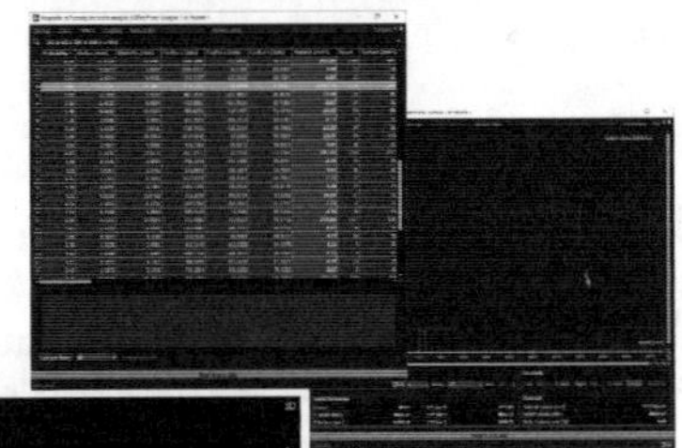

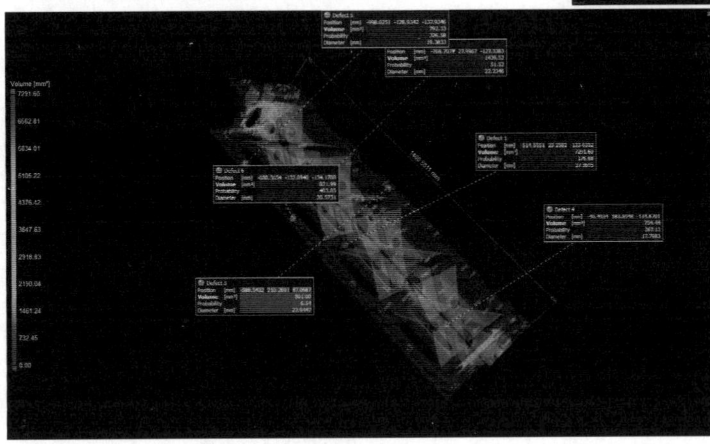

- Results can clearly identify internal porosity and cracks where 2D x-rays show inconclusive or no results
- Results can be quantified by volume and location for each internal void or pore
- Parts no longer have to be cut up to locate internal porosity

LARGE FORMAT METAL PARTS – REVERSE ENGINEERING

Development of a CAD file with internal and external geometry from a CT dataset. After a part has been CT scanned, internal and external data are extracted from the results. Data can be currently exported in the following formats: STL, WRL, TXT, PLY, OBJ.

Additional Options:

STL To STEP
Point Cloud to Surface Files
FEA Analysis/Mold Flow/
Magma

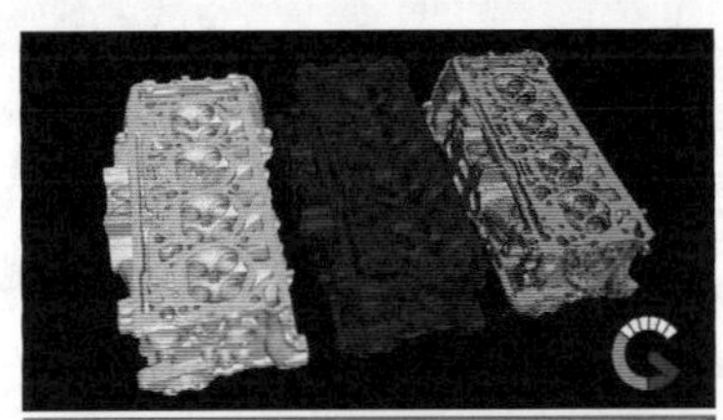

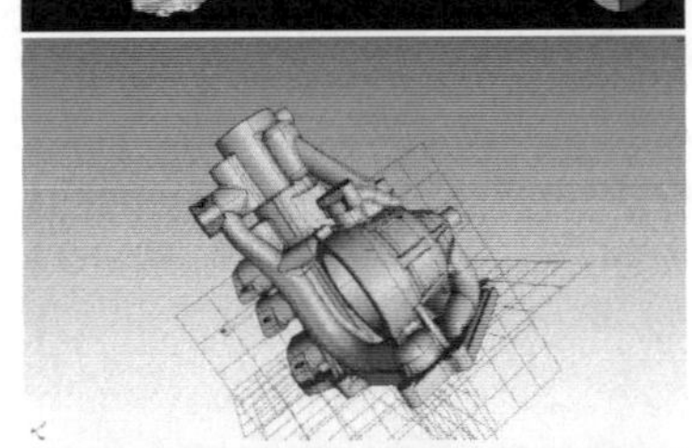

LARGE FORMAT METAL PARTS – AM PARTS

A quick check of a additively manufactured part can determine if there is unfused powder pockets, delamination of layers, keyholing and geometry of internal features.
See images to the right of trapped unfused powder

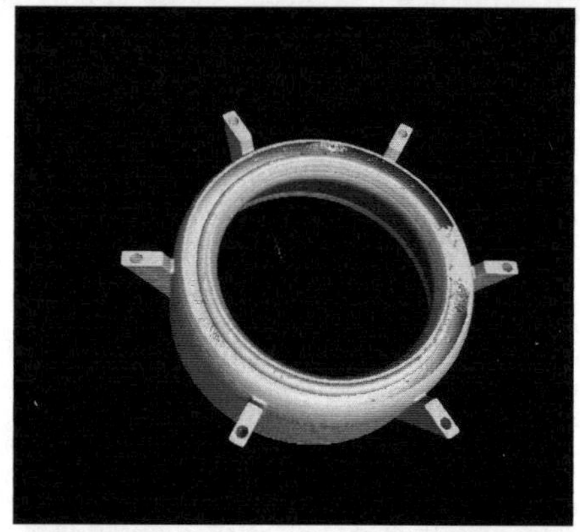

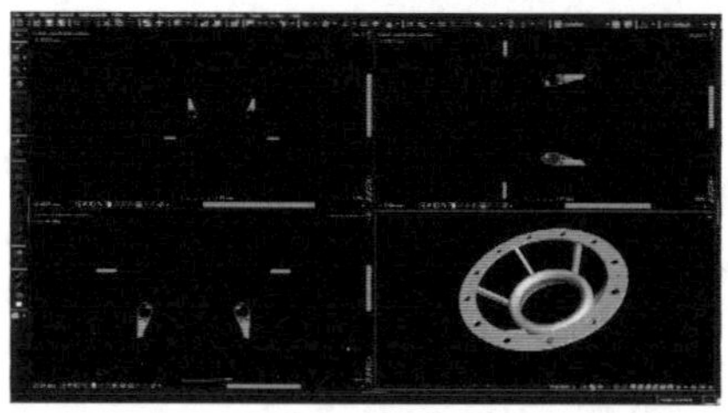

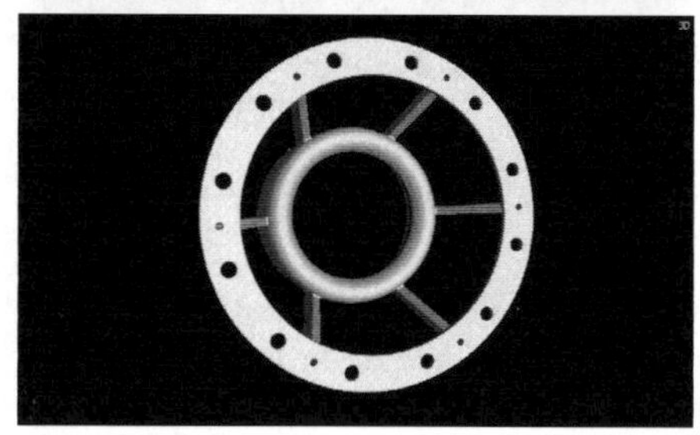

LARGE FORMAT AM PARTS – INCONEL - VOIDS

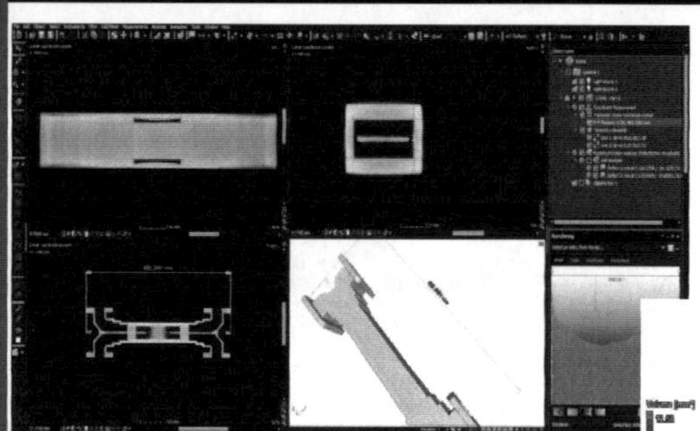

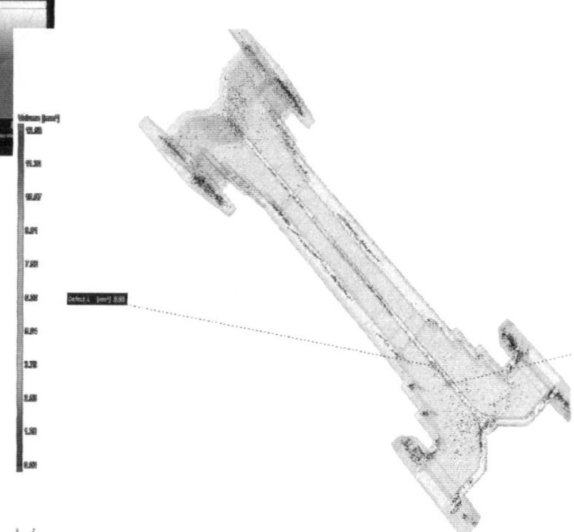

SUMMARY

Advantage of using a linear accelerators for CT

- Fast acquisition and scan time.
- Large field of view
- Shorter scan times equals less wear and tear on expensive equipment
- More versatility in scanning techniques versus LDAs
- Clean differentiation between mixed materials and complex assemblies.
- High level of resolution and accuracy
- Large higher density and low density materials that cannot be scanned using traditional tubes <1MeV

Special thanks to Tony Melton from Industrial Imaging Solutions and his ZxAcquire DDA Software.

Etch Defect Detection by Faster R-CNN

P. Deschênes-Labrie, F. Qiang, K. Haddad, L. Duc Thien,
N. Carl, J. Alejandro, W. Yizhe, X. Maldague

CINDE
CANADIAN INSTITUTE FOR NON-DESTRUCTIVE EVALUATION

NDT in Canada
NDTC 2018
Canada's NDT Conference
June 19 - 21
Halifax Convention Centre
Halifax, Nova Scotia

Etch Defect Detection by Faster R-CNN

- **Outline:**

1) Surface defect: Etch defect
2) Deep learning from CNN to Faster R-CNN
3) Results
4) Pros & Cons
5)Recommendation
6) Conclusion

Etch Defect Detection by Faster R-CNN

Machine learning on vision

- **Advantages**
  - Effective at solving problems: bounding boxes
  - More data equals more accuracy and more stability compared to manual testing
  - Faster to implement than classic computer vision
- **Challenges**
  - Dataset
  - Etches is different: pixel level -> harder to detect

Etch Defect Detection by Faster R-CNN

Preprocessing phase

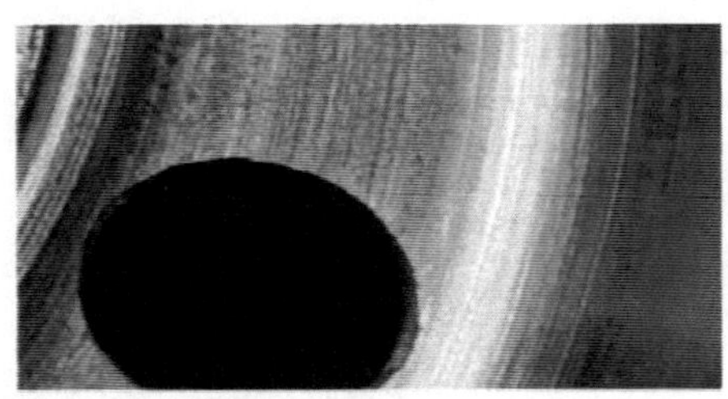

Defects may be clearly observed or ambiguous

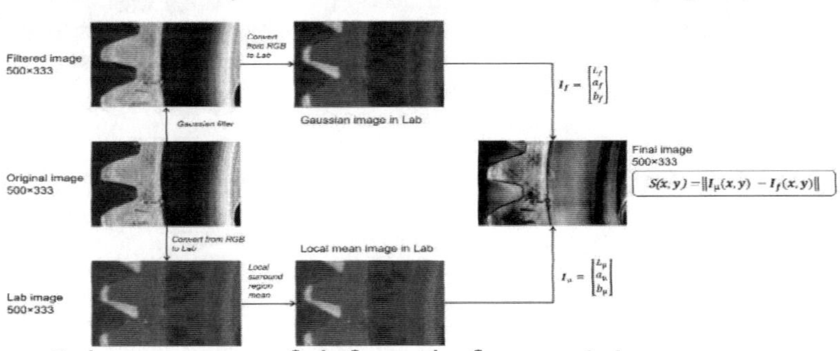

Enhancement of defects before training

Etch Defect Detection by Faster R-CNN

- R-CNN

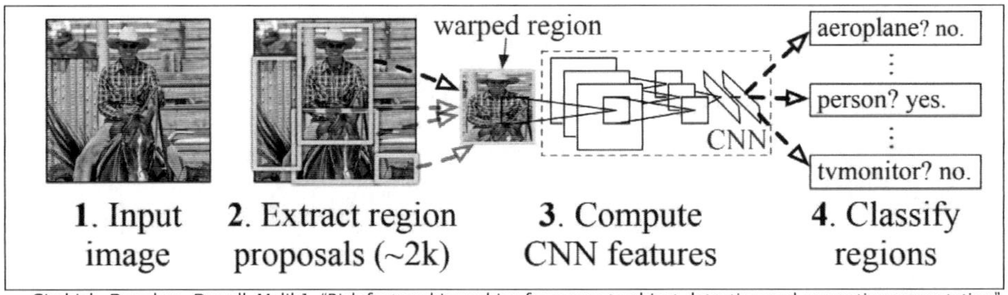

Girshick, Donahue, Darrell, Malik1, "Rich feature hierarchies for accurate object detection and semantic segmentation"

- Faster R-CNN

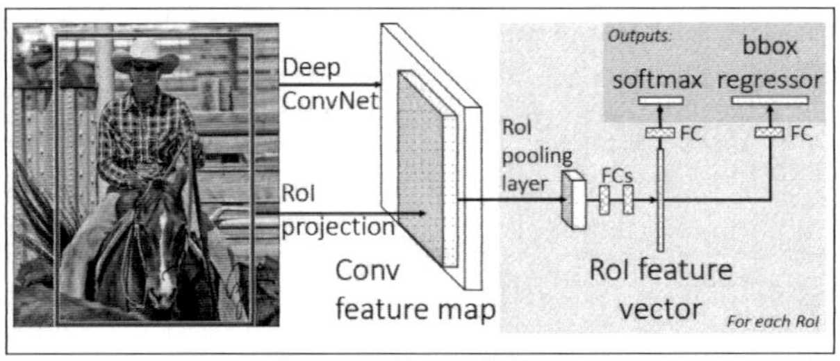

Ross Girshick, "Fast R-CNN"

Etch Defect Detection by Faster R-CNN

- Dataset building

Dataset:

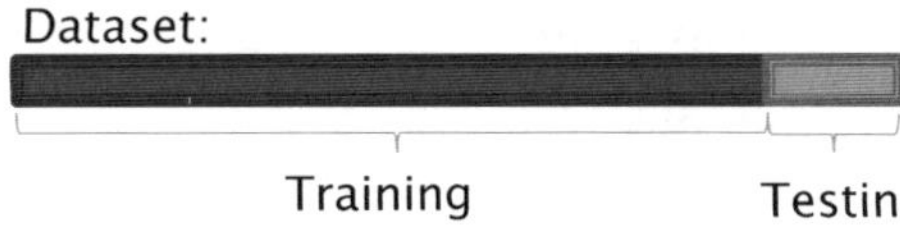

Etch Defect Detection by Faster R-CNN

- **Defect detection proposals**

Region proposals

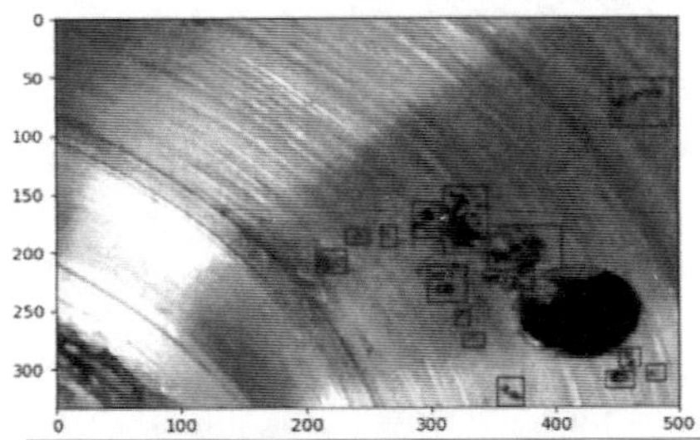

- Building from 4 cameras
- Hand marking one type of defect using bounding box
- Same part used for all the dataset

Programs

- Anaconda
- Package : DeepLearning Caffe
- Pascal VOC
- Python - PyTorch & CUDA

Etch Defect Detection by Faster R-CNN

- Results

Etch Defect Detection by Faster R-CNN

- Results

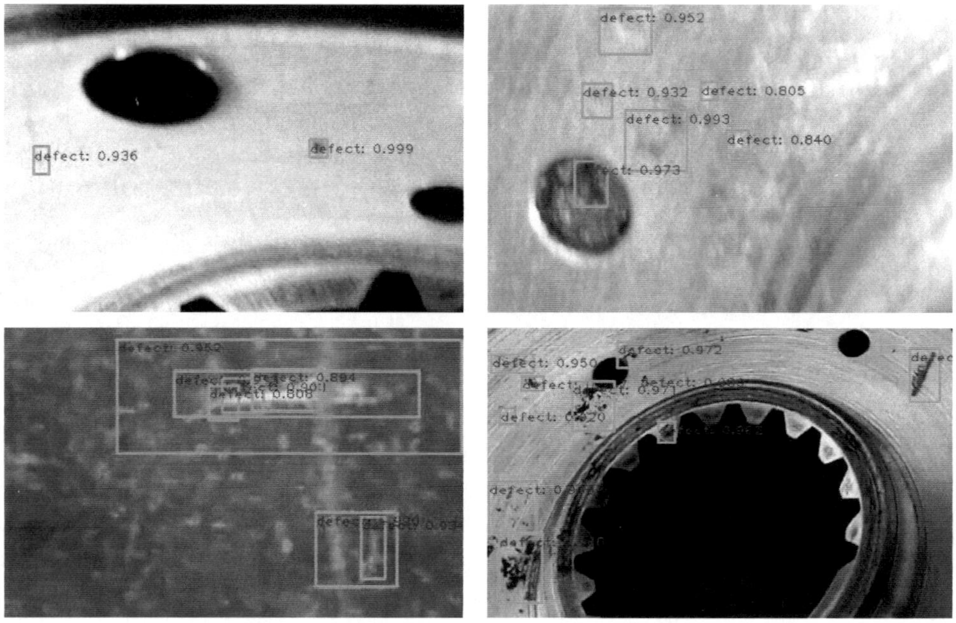

NDT in Canada 2018 | June 19-21 | Halifax, NS

Etch Defect Detection by Faster R-CNN

- Results

NDT in Canada 2018 | June 19-21 | Halifax, NS

Etch Defect Detection by Faster R-CNN

- Results: Confusion Matrix

Class	Defect	Non-defect
Defect	889	162
Non-defect	52	200

Table 1: Class Mark List

$$Accuracy = \frac{\texttt{Correctly classified regions}}{\texttt{Total number of evaluated regions}} = 83.57\%$$

Etch Defect Detection by Faster R-CNN

- Results: Hit/Miss/Fail

Training 1

Training 2

Etch Defect Detection by Faster R-CNN

Results: Hit/Miss/Fail

Training	1				2			
Picture	Hit	Miss	Fail	% Defect Detection	Hit	Miss	Fail	% Defect Detection
Test 1	6	0	1	100.00	5	0	0	100.00
Test 21	6	1	2	85.71	5	1	0	83.33
Test 30	7	0	3	100.00	7	0	3	100.00
Test 41	10	2	0	83.33	10	2	0	83.33
Test 50	2	0	5	100.00	2	0	6	100.00
Test 59	5	0	1	100.00	5	0	0	100.00
Test 61	7	3	3	70.00	9	1	1	90.00
Test 70	7	0	3	100.00	8	0	2	100.00
Test 79	8	3	2	72.73	8	2	2	80.00
Test 91	4	0	3	100.00	4	0	2	100.00
Test 99	1	0	1	100.00	1	0	1	100.00
Test 108	4	1	5	80.00	3	1	1	75.00
Test 116	8	3	2	72.73	6	3	1	66.67
Test 125	5	4	3	55.56	8	2	2	80.00
Test 133	6	0	2	100.00	3	0	0	100.00
Test 139	9	0	1	100.00	10	0	0	100.00
Test 150	5	1	0	83.33	4	1	0	80.00
Test 160	1	0	1	100.00	1	0	0	100.00
Test 165	6	1	4	85.71	7	1	2	87.50
Test 190	5	1	3	83.33	6	0	0	100.00
TOTAL				**84.85**				**88.89**

- **Hit**: Defect spotted
- **Miss**: Defect not spotted
- **Fail**: Bounding box on nothing

NDT in Canada 2018 | June 19-21 | Halifax, NS

Etch Defect Detection by Faster R-CNN

Pros & Cons

- Pros
  - Robust through blur & lighting
  - Possibility to do continuous learning
  - Doesn't require any special setup*

- Cons
  - Need to build initial database
  - Needs a lot of pictures
  - Initial parameters need tuning
  - Training time and computer resources

NDT in Canada 2018 | June 19-21 | Halifax, NS

Etch Defect Detection by Faster R-CNN

Recommendations

To gain reliable results, we need to have :

- A lot of pictures for the training

  Approach works by seeing different scenarios

- To take pictures of all the parts/defects

  The algorithm is very good for what it trained for

NDT in Canada 2018 | June 19-21 | Halifax, NS

Etch Defect Detection by Faster R-CNN

Conclusions

- The proposed Faster R-CNN approach can detect defects and may be further improved to gain higher accuracy.

- It helps to improve the inspection process for the personal by showing them where possible defects are.

NDT in Canada 2018 | June 19-21 | Halifax, NS

Etch Defect Detection by Faster R-CNN

Acknowledgements

- Pratt & Whitney Canada is deeply thanked for its help and assistance through this project.

Etch Defect Detection by Faster R-CNN

THANK YOU
FOR YOUR ATTENTION!

Increased process safety and efficiency through ADR

Automated Defect Recognition and automation in X-ray inspection

NDT in Canada
NDTC 2018
Canada's NDT Conference

June 19 - 21
Halifax Convention Centre
Halifax, Nova Scotia

CINDE
CANADIAN INSTITUTE FOR NON-DESTRUCTIVE EVALUATION

About VisiConsult

- Manufacturer of customized X-ray systems
- Design, manufacturing, software and applications
- 70 employees worldwide
- Located in Northern Germany: Stockelsdorf
- Family owned business (second generation)
- ISO9001 certified processes
- US subsidiary based in Atlanta

NDT in Canada 2018 | June 19-21 | Halifax, NS

▸ Parameterization

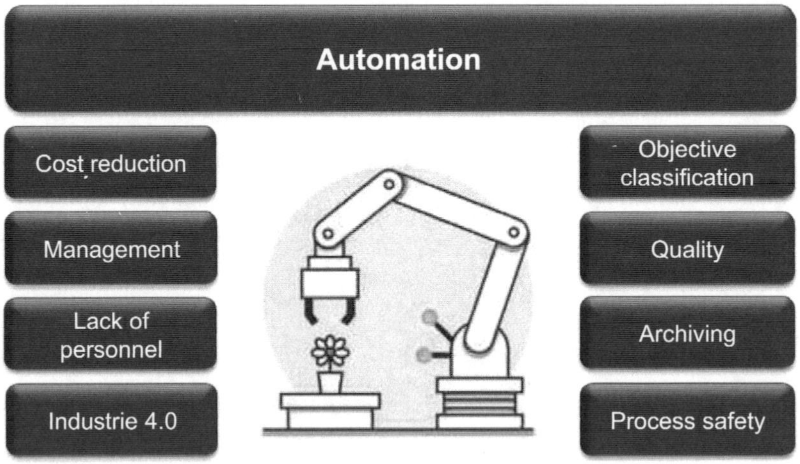

▸ ADR results

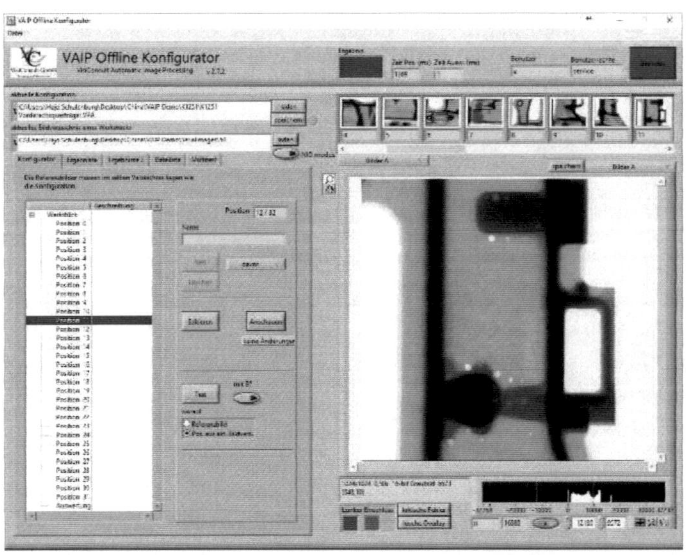

‣ ADR results

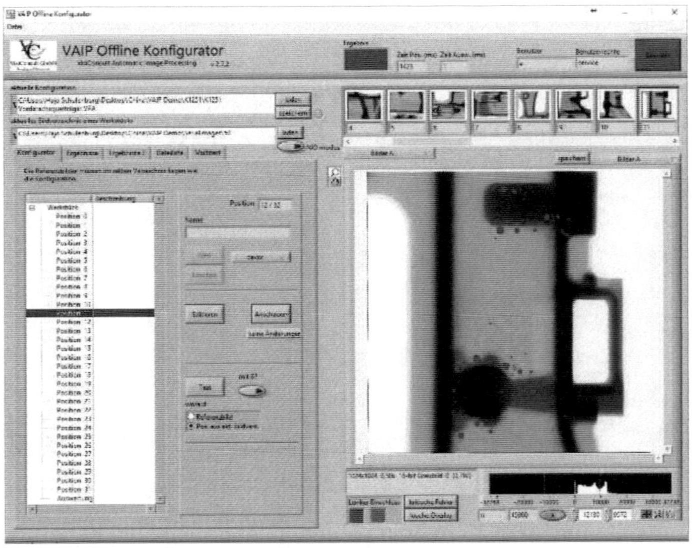

NDT in Canada 2018 | June 19-21 | Halifax, NS

‣ ADR configuration - ROIs

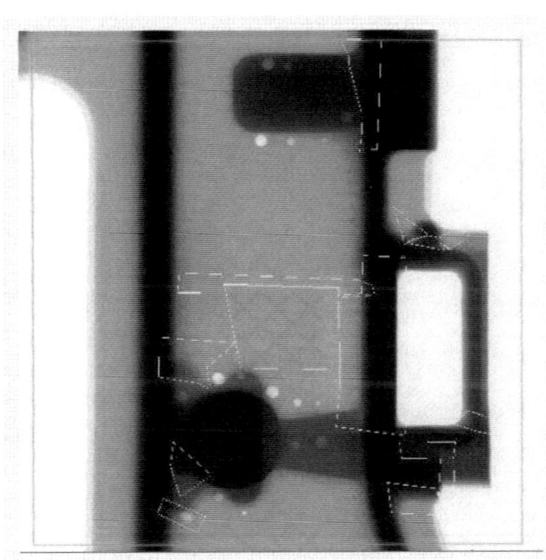

NDT in Canada 2018 | June 19-21 | Halifax, NS

Parameterization

- Metrics:
  - Defect size
  - Defect size per area
  - Minimum surface distance
  - Minimum defect distance
  - Etc.

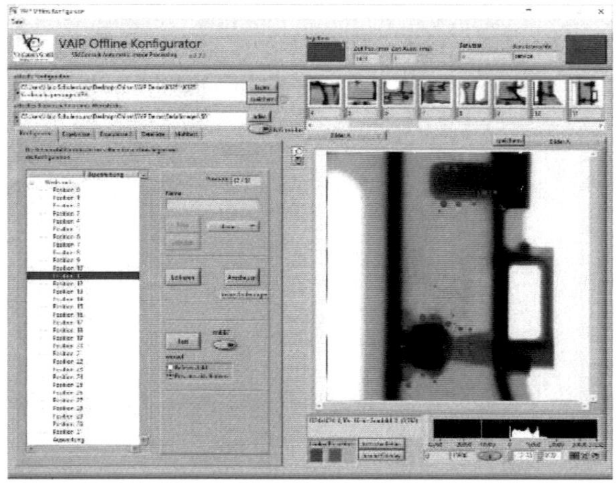

Results

- Export:
  - Defect list
  - Decision
  - Defect position
  - Defect shape
  - Report
  - Etc.

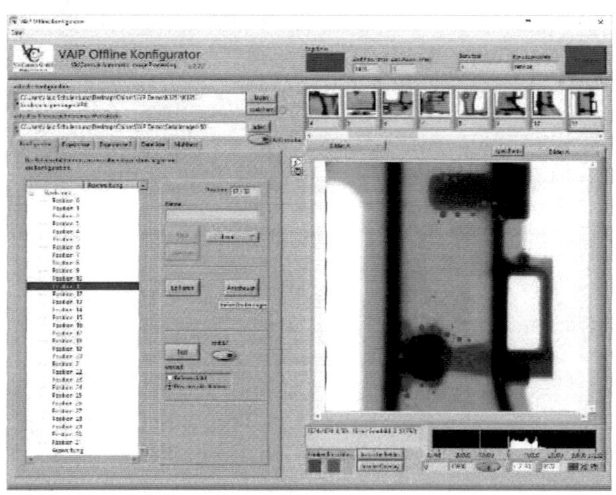

ADR configuration - ROIs

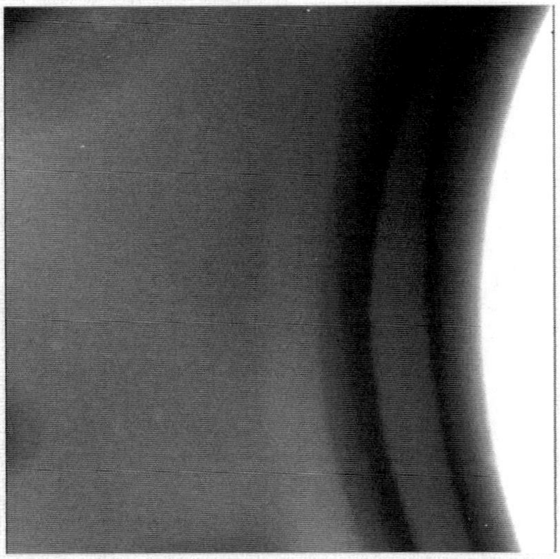

NDT in Canada 2018 | June 19-21 | Halifax, NS

ADR configuration - ROIs

NDT in Canada 2018 | June 19-21 | Halifax, NS

ADR configuration – ROIs

ADR configuration – ROIs

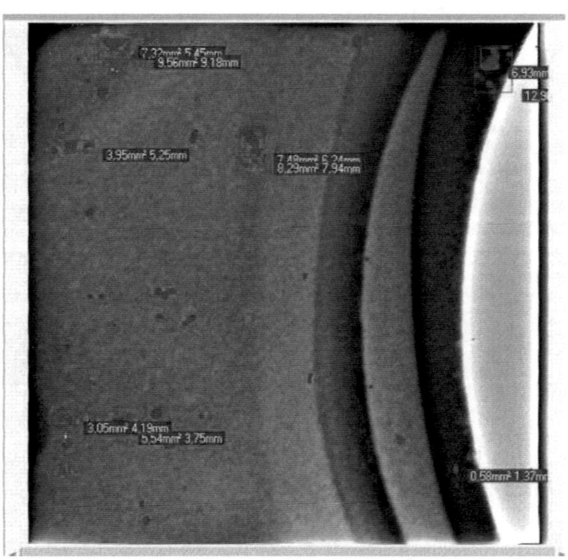

ADR configuration - ROIs

NDT in Canada 2018 | June 19-21 | Halifax, NS

ADR configuration - ROIs

NDT in Canada 2018 | June 19-21 | Halifax, NS

ADR configuration - ROIs

ADR configuration - ROIs

False positive reduction

NDT in Canada 2018 | June 19-21 | Halifax, NS

False positive reduction

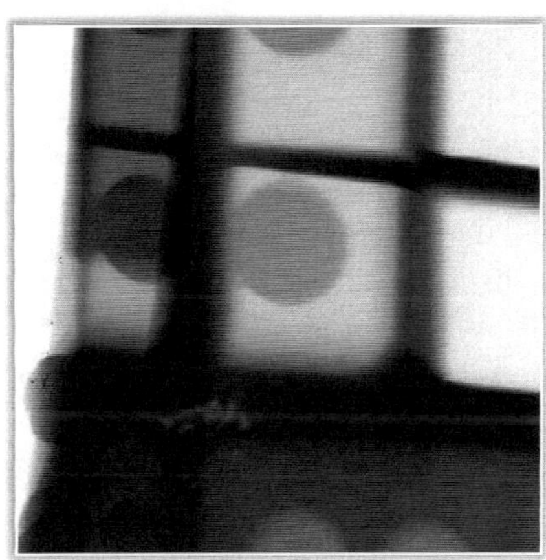

NDT in Canada 2018 | June 19-21 | Halifax, NS

▸ False positive reduction

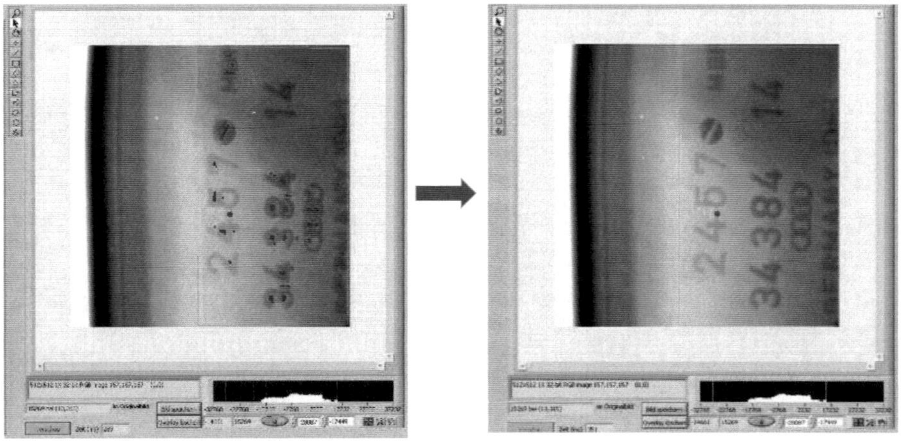

▸ Image registration

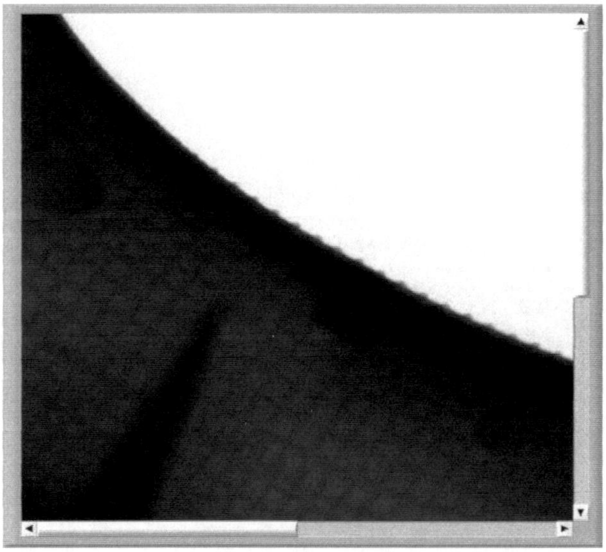

Image registration

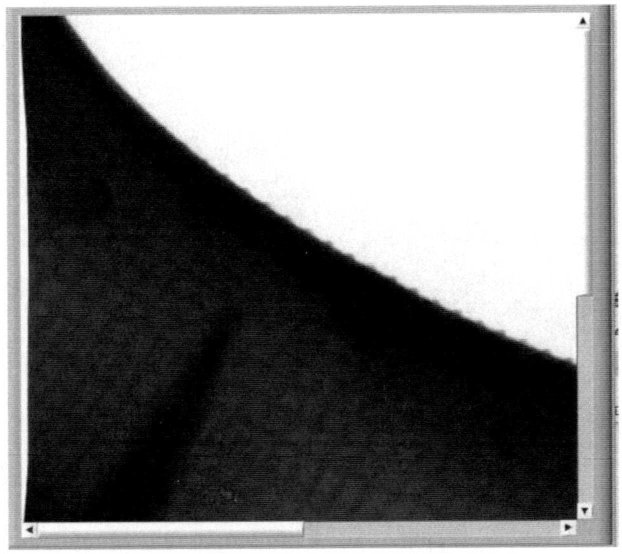

NDT in Canada 2018 | June 19-21 | Halifax, NS

Artificial flaws - verification

NDT in Canada 2018 | June 19-21 | Halifax, NS

OCR - Text Recognition

NDT in Canada 2018 | June 19-21 | Halifax, NS

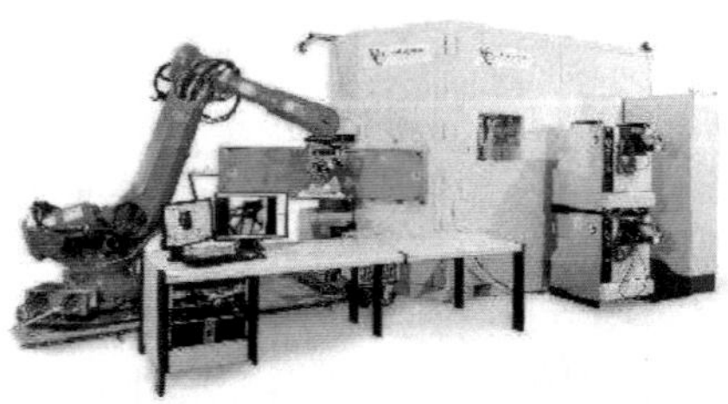

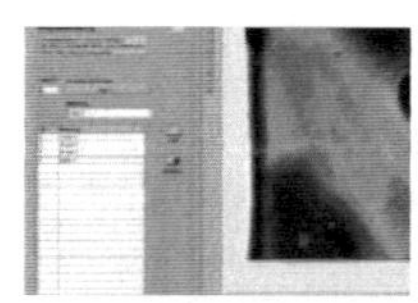

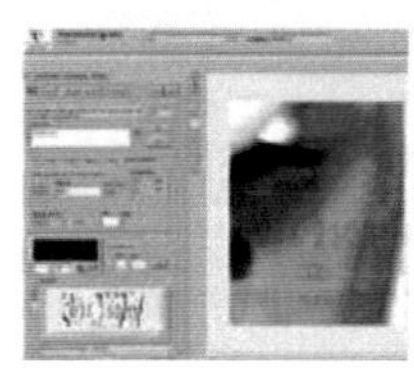

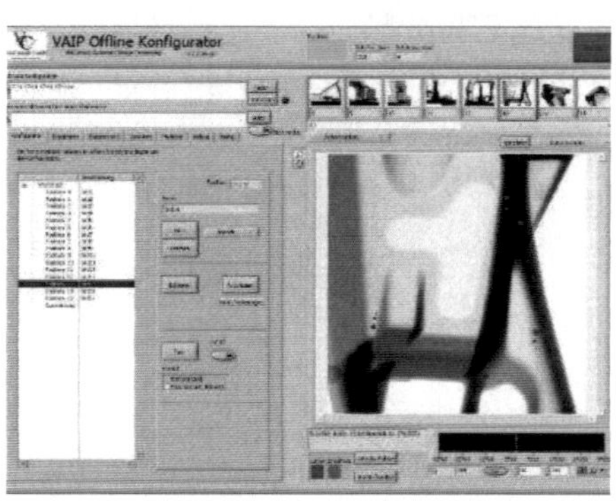

- Porosities
- Inclusions
- Geometry
- Serialnumbers

- Approved by BMW
- Fastest ADR system worldwide
- 24/7 operation
- No human operator

NDT in Canada 2018 | June 19-21 | Halifax, NS

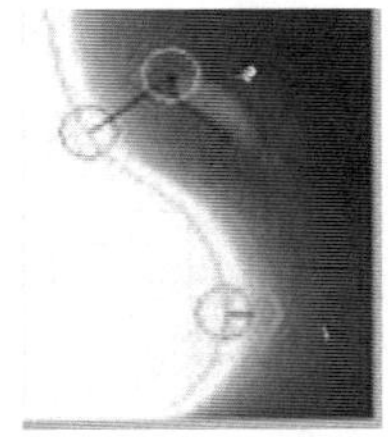

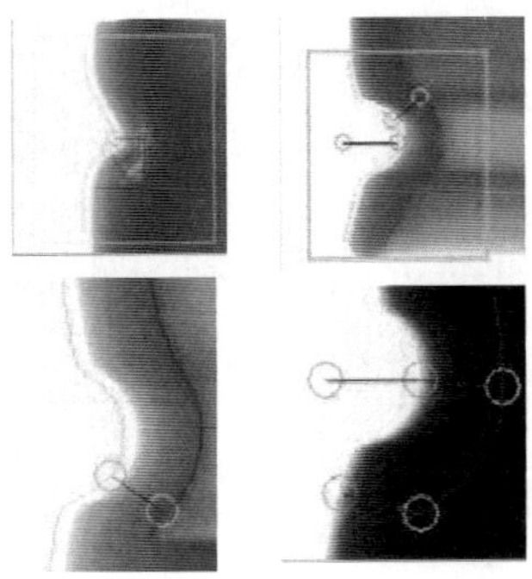

- Geometry
- Density
- Thickness
- Missing material and pores

- For airbag igniters
- Automation of DR cabinets
- 24/7 operation
- No human operator

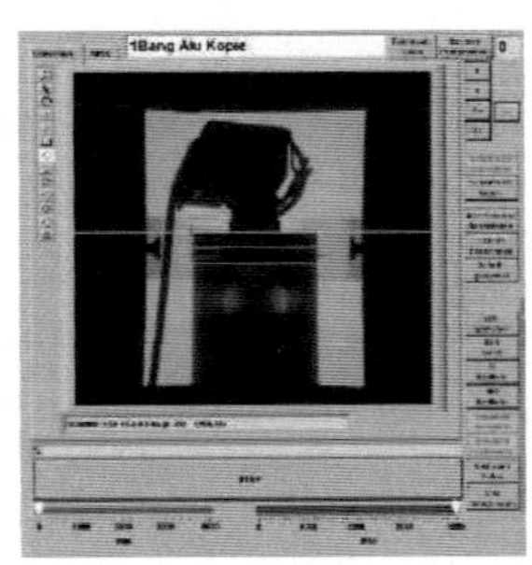

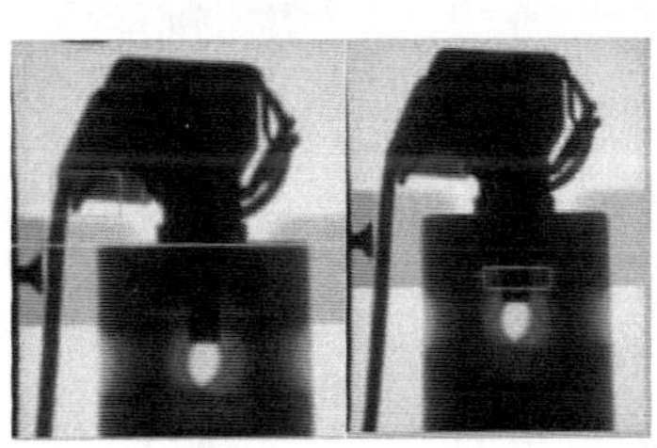

- Foreign Material
- Missing components
- Wrong component
- Destiny
- Geometry

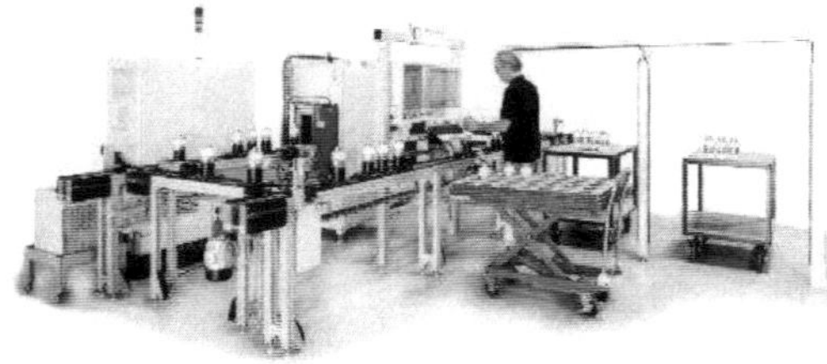

SIEMENS

- Foreign Material
- Missing components
- Wrong component
- Geometry

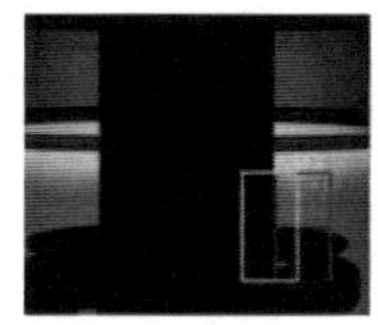

- Fully automatic
- Hundreds of checks
- Rotating device
- High voltage switches

Computed Tomography ADR

Aluminum Casting

Reconstructed slice   Porosity Analysis

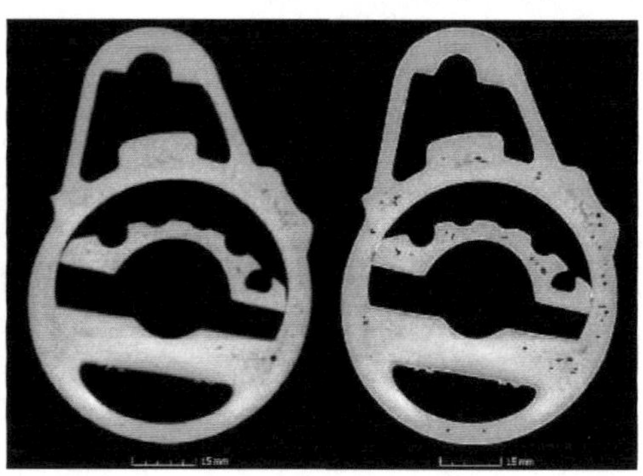

Thanks for your attention

VisiConsult
X-ray Systems & Solution GmbH
+49 (0) 451 2902860
info@visiconsult.de

NDT in Canada 2018 | June 19-21 | Halifax, NS

Installed Inspection Sensors & Cloud-based Software for Continuous Corrosion Monitoring and Advanced Analytics

Thaer ALKHOURI, Tara MERRY,

Phone: 1 403 462 2318, E-mail: thaer.al-khoury@bhge.com, Tara.Merry@bhge.com
Baker Hughes, a GE company - Inspection Technologies, Calgary, Canada

Abstract

Corrosion and erosion cost industries billions of dollars every year in, "lost production, unplanned downtime. equipment failure, high maintenance repair costs, imposed fines or compensation costs." The president of NACE International, the U.S. professional technical society dedicated to reducing corrosion, said, "The cost of corrosion is a waste of resources and investments." He then went on to state that the direct cost of corrosion is more than 3% of the U.S. Gross Domestic Product (GDP). That is more than $300 billion annually in the United States alone. That global cost of corrosion is $2.5 trillion dollars, and represents ~6% of global GDP.

However, both NACE.org and NASA studies estimate that companies can save between 15% to 35% of annual corrosion-related costs through optimum corrosion management practices. That means working together to address this problem through a variety of means could result in $330-$550 billion-dollar savings globally.

The most frequent scenario of plant failure is wall thinning due to erosion and corrosion. Conventional inspection "NDT" methods are deployed in order to measure, map, and monitor the thinning rate. These inspections create significant cost, can result in downtime, and present safety risks, especially those that required in difficult-to-access locations. In addition, each technology has limitations, and often human factors affect the accuracy and accountability of data for long-term analyses or corrosion trending.

This paper aims to present the value of deploying continuous wireless corrosion monitoring of interior piping wall thickness to mitigate or control the effects of corrosion and increase confidence on corrosion data. In addition, it will explain the differentiations between conventional inspection "NDT", intrusive methods, and continuous corrosion monitoring. Lastly, the paper will demonstrate how implementing installed inspection sensors can decrease inspection costs, improve asset performance, and optimize process/operations

Keywords: Corrosion Monitoring, Digital Inspection, Asset integrity, Corrosion Management

1. Introduction

In order to comply with industry regulations, protect employees, assets, and the environment, and avoid equipment failures and unplanned shutdown, industries are spending billions of dollars. The goal of this type of investment is to enable a better understanding about the corrosion challenges and overall integrity picture at a site. Corrosion control remains a particularly challenging phenomenon that need will continue to evolve. Continuous monitoring plays a vital part of that evolution

Why Corrosion Management?

Aging:

- Most assets are aged, "some of which now have been in operation for double the original design life"
- 60% of worlds pipelines are ~40 years old, with an elevated risk of failure

Cost

- Inspections for corrosion are expensive, require specialized skill sets, and are often required in difficult-to-access locations
- Corrosion can result in unplanned downtime, maintenance, and costly repairs
- Unanticipated piping wall loss or acceleration in corrosion rate can result in downtime driving the loss of production, and lost revenue

Safety

- Dangerous site access "hard-to reach, high & low temperature, confined space entry"
- Hazardous environments: radiation, extreme temperatures, highly combustible gas, toxic or acidic vapor

Reputation / Image

- Multi-million-dollar fines
- Image damage to company when accidents happen

2. Corrosion Management: Conventional Methods

2.1 Traditional ultrasonic inspection

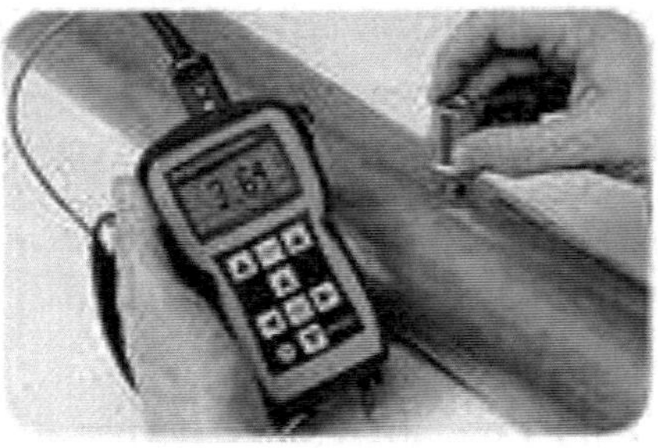

Regularly scheduled inspections will validate corrosion rates and allow engineers and operators to better plan for maintenance situations. The number and location of Thickness Monitoring

(TMLs) that needs to be a tested will depend on the required confidence we are looking for. More locations should be used to monitor higher risk locations, driving cost & inspection time up.

Additional challenges:

- Measurements are point-in-time, and often infrequent, rather than continuous: not suitable for short-term corrosion rates, need for more and quicker data with higher quality
- Trained and certified inspector is required
- Gaining access to the site, including scaffoldings, rope access, and insulation removal are time-consuming and expensive
- Hazardous environments, and access during difficult conditions
- Location repeatability … ensuring that the exact same spot on the piping is going to be measured over a period of time
- Performance and accuracy factors: equipment used, variation on the equipment calibration, surface roughness, coupling technique, curvature of the test piece, velocity variations, temperature
- Human factor, i.e. expertise of the operator, actual performance of the testing in the field
- Equipment & hardware costs

2.2 Intrusive methods

Intrusive methods like corrosion coupons attempt to manage a target corrosion rates set by the producer

- May be based upon corrosion allowance and asset lifetime
- May use NACE defined corrosion rates

RP0775-2005

Table 2: Qualitative Categorization of Carbon Steel Corrosion Rates for Oil Production Systems

	Average Corrosion Rate		Maximum Pitting Rate (See Paragraph 2.5)	
	mm/y[(A)]	mpy[(B)]	mm/y	mpy
Low	<0.025	<1.0	<0.13	<5.0
Moderate	0.025-0.12	1.0-4.9	0.13-0.20	5.0-7.9
High	0.13-0.25	5.0-10	0.21-0.38	8.0-15
Severe	>0.25	>10	>0.38	>15

[(A)] mm/y = millimeters per year
[(B)] mpy = mils per year

Intrusive methods most- commonly used include:

- Weight loss coupons
- LPR (Linear Polarization Resistance)
- ER (Electrical Resistance)
- Metal ion analysis (iron counts)

Each method has its own advantages and disadvantages, including those highlighted as-follows:

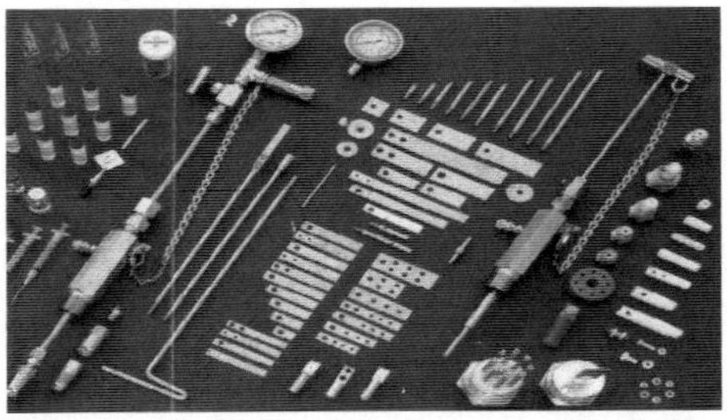

- Indirect measurement: these tools provide data on the corrosion rate of the fluid, but do not provide accurate data on the wall thickness. Measurement is carried out on a shining piece of metal, meaning that different materials experience different levels of corrosion.
- Flow changes: these tools could create some change (could be small) on the internal flow that which could generate relatively different corrosivity conditions to the rest of the process, and data could be highly variable
- Installation, maintenance, download data and retrieval would require access, driving higher risks, lost time, and cost.
- Solid deposit could generate misleading results and short-circuit mechanism of reading.
- Allow for global evaluation of the corrosion process but doesn't indicate where corrosion occurs
- Would measure corrosivity for the position it is in (not area)
- Could be placed at locations where they are easy to install / collect data, not necessarily where corrosion is likely to occur

"In order to avoid undesirable effects, a chemical injection program is almost always coupled with a monitoring and inspection program …"

3. Implementation of New Technologies

3.1 Implementation of Digital Technologies & Wireless Connectivity

Digital technologies and wireless connectivity are revolutioning all industries, and applied across a wide range of applications. They are creating new innovations, and shifting business models to create new demands.

Cloud technology has enabled data availability, integration, storage without limits, and computation capacity. Inspection data is no longer a representation of a single point in time, but rather becomes an informative and transformational business tool. Cloud technologies enable easier access, real-time updates, and more effective data storing, integrating and sharing across the business network. Today, less than 10% of the data in oil and gas companies is effectively used. The cloud unlocks data so that more-informed decisions can be made across the enterprise.

With the acceptance of change come significant improvements to the world of inspection: centralized, cloud-based technology, sensor and monitoring technologies, drone and robotics technologies, and remote collaboration. The results of these innovative technologies are increased safety, increased productivity, improved data quality, and overall improved asset integrity, with reduced downtime, cost, and risk.

Over the past several years, sensors have become notably more powerful and cost effective, enabling a broader scale and depth of information to be continuously available. Installed on-site and connecting directly to the cloud, sensors can provide real-time visibility across assets through a simplified, central view of inspection data, including thermal and thickness measurements. Instead of periodic, point-in-time inspections that leave windows in which a failure can occur, and provide a limited view of corrosion risk, organizations can continuously "inspect", detect actual problems, and predict potential areas of concern. This enables more proactive asset management, keeping operations both productive and safe.

3.2 Implementation of Dry-Coupled Ultrasonic Sensors

Ultrasonic sensor, applied directly on the pipe provide 0° pulse-echo ultrasound measurements of the wall thickness.

Advantages to this type of sensor include:

- Footprint of only 1x1", meaning a higher density of sensors can be achieved
- Low profile of only 16mm, meaning the sensors can easily be placed under insulation
- Accuracy measurements of the wall thickness, with a resolution of 0.1 mm or better.
- Testing has shown that precise measurements can be obtained with a repeatability margin of 0.02 mm.
- Sensors are ATEX certified
- SolGel Ceramic Technology: Dry coupled, IP67, meaning sensors can be installed without the need for a couplant
- Sensors have minimum of 10 years lifespan with an estimated 15-20 years lifespan without SNR loss, even with temperature cycles
- Thermocouples are installed with sensors, meaning each A-scan is paired with a temperature measurement. This data is saved in a database and undergoes post processing to calculate the wall thickness.
- The sensors are clamped on as a non-intrusive method, no welding is required to install, hence there is no shut down required.

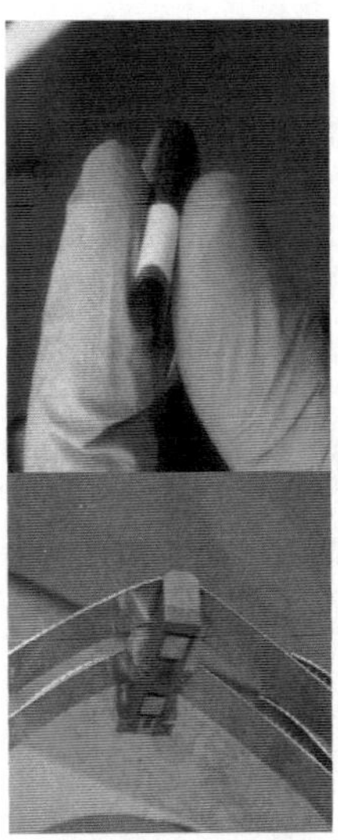

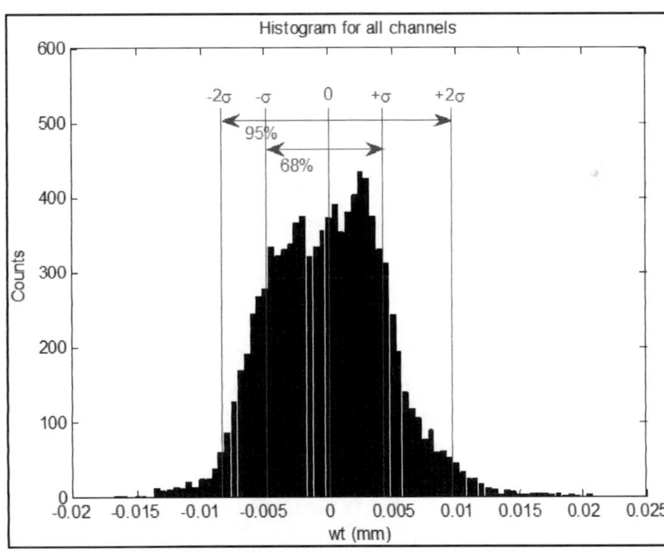

A histogram showing the distribution of 40k measurements with 95% of measurements fall within a margin of 0.02 mm.

Sensors can be attached using a variety of methods and configurations on straight pipes, elbows, headers on different orientation, corrosion mechanism in combination with the most critical area of a certain asset will drive the probe locations

4.Wireless Communication and Data Acquisition

The goal of monitoring is to provide insights enabling experts to get smarter about their assets. This allows them to better-predict how they will perform, make informed decision about when to repair or replace, so they can maximize the value of investments while simultaneously reducing the risk of catastrophic failures.

This marriage of the latest digital technologies, wireless connectivity, and dry-coupled ultrasonic sensors can offer real-time thickness measurement, and the ability to perform advanced data analytics. This shift enables improved asset integrity management, and simultaneously opens the door for innovative ways to utilize this data to optimize plant and process performance.

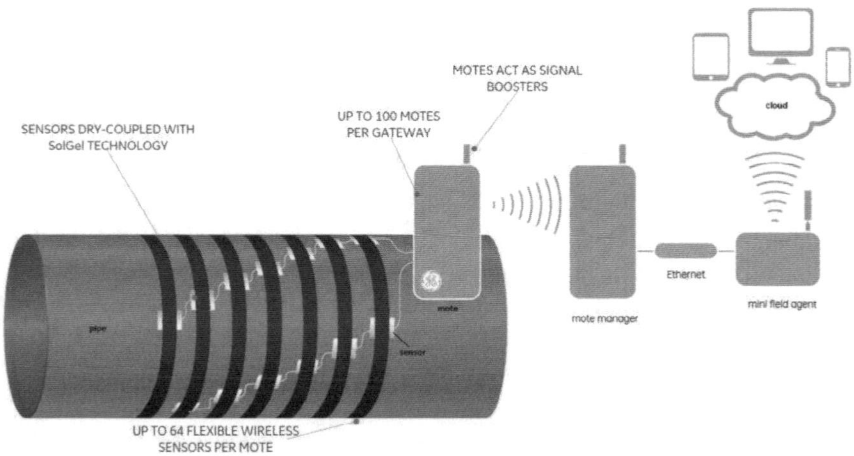

Each probe is controlled by a battery-powered controller box called the 'mote'. The mote capable of serving up to 64 probes. Each individual probe is connected to the mote by a single coax cable and powered by battery with a minimum of 5 years battery life, guaranteed for 64 sensors. The battery life has a potential life well-beyond 5 years, dependent upon the volume of sensors and frequency of readings.

The Mote manager gathers and transfers measurements from multiple motes to the mini field agent. Up to 100 motes can communicate to the mote manager, making a mesh network. Within a mesh network, every device can communicate to the mote manager either directly or via one or more other devices in the network.

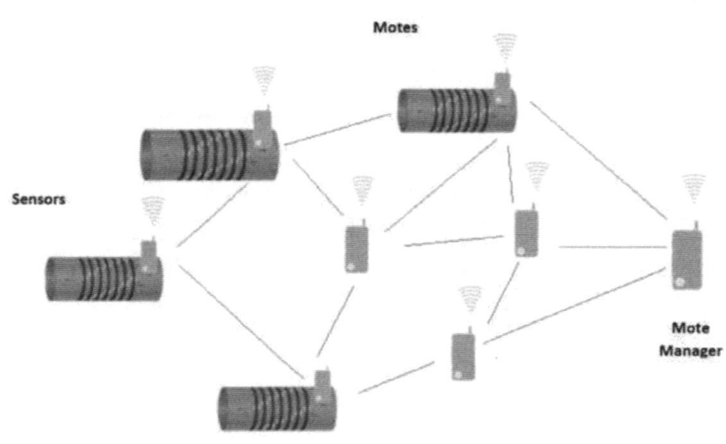

Additional motes can be added to cover the mesh network

The mote managers are powered via Power Over Ethernet, and hard wired with an Ethernet cable to the Mini Field Agent.

The mini field agent orders the motes through the mote manager to wake up and take measurements. The mini field agent also encrypts the data received and sends the data to the Predix Cloud essentially acting as a gateway between the hardware and the software.

The system operates a private, segregated wireless mesh network for sensor data collection. This network is managed by the mote manager, which uploads processed sensor data to the GE Predix cloud via ethernet. It is therefore not required to be integrated into existing wireless deployments, and does not interfere with site standards adoptions.

5. Data Analysis Software

The sensors are continuously collecting interior piping wall thickness and temperature readings, and feeding them into the cloud-based software. Users can benefit from more reliable inspection data, and then start to trend and predict failures due to corrosion and erosion.

Software benefits include:

- Geo-spatial view allows you to see your assets in real-time
- Plant summary highlights the areas of concern by level of criticality, with a summary trend
- Stores the full A-scan ultrasonic testing data, as well as the calculated thicknesses
- Data analysis, reporting and exporting
- Remote control, automatic scheduling
- Diagnostic review of remote sensors, the motes, and other components so that it's doing its job of monitoring your production assets.
- Temperature adjusted thickness readings

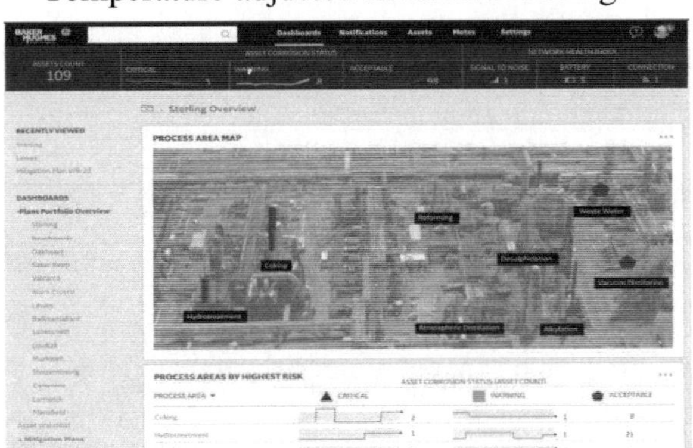

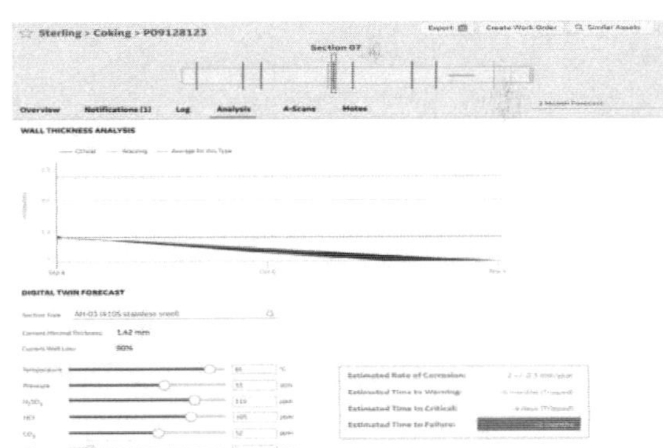

This information will improve the reliability of inspection data, and enables users to trend and predict piping failures due to corrosion. The result is enables asset owners to extend the life of piping assets across a facility.

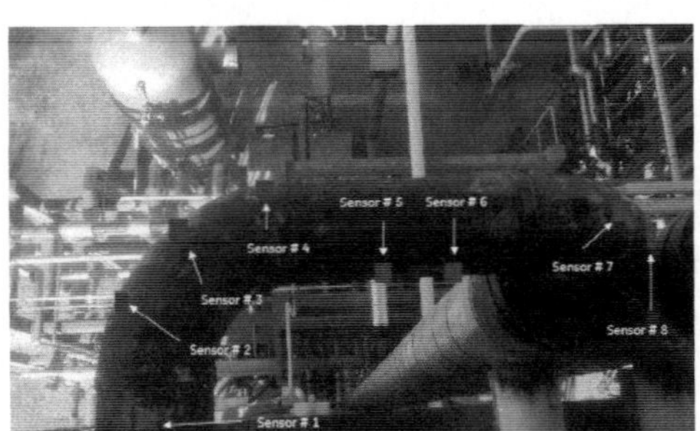

6. Conclusions

Continuous corrosion management solutions move operations beyond manual inspection to continuously connect assets, data, and people. The technology provides the actionable intelligence needed to access data from virtually anywhere. It allows users to forecast remaining asset life, predict future corrosion-related risks, manage related risk, and proactively maintain systems.

Results from continuous monitoring:

Decrease inspection costs:
25% savings potential or $250K/year for average refinery inspection budget = $1M/year

- Offset points with high cost-of-collection
- Optimize with better decisions from continuous data
- Identify critical points that are unmonitored today

Improve asset performance:
15% asset life extension potential, through better maintenance and the identification of equipment that may need modification to its operating profile or chemical treatments.

Optimize process/operations:
15% improvement in process efficiency

- Extend run-times
- Optimize corrosion inhibition
- Make better decisions through predictive analytics

References

1. GE Measurement & Control, New! Predictive Corrosion Management
2. NACE Standard RP0775-2005
3. Gysbers, A.C. "A Discussion on the Piping Thickness Management Process: Determining Corrosion Monitoring Locations." Inspectioneering Journal, vol. 18, no. 6, 2012.
4. Evolution in Digital Inspection Technologies Mark Rosenberg – GE digital, 2017
5. GE Measurement & Control, Predictive Corrosion Management literature, 2017
6. "Corrosion Costs and Preventive Strategies in the United States", Supplement to Materials Performance, NACE International, Houston, TX, July 2002, p. 3.

Structural Mechanical Simulation on Voxel Data

Roger WENDE[1], Johannes FIERES, Christof REINHART

[1]Volume Graphics, Inc., 415 Minuet Lane, Suite C, Charlotte, NC 28217, USA
e-mail: wende@volumegraphics.com

Abstract
As Computed Tomography (CT) matures in the industrial market, the need for further analysis tools for voxel data has increased. Today, it is common for engineers to use a Von Mises analysis to express local stresses induced on test samples. The classic method for determining when a material fails is a *destructive tension test*. So, what can CT bring to this already proven method for calculating values such as a Von Mises analysis and identifying hotspots where failure may occur? CT data reveals not only the surface information of a part, but also the internal features of a manufactured part, e.g. the defects (voids) that occur in manufacturing. Combining the geometry surface information, which can be a complex design in additive manufacturing with defects introduced, we can better analyze the tensile strength of an actual part non-destructively, using structural mechanical simulation. By using CT data only, one can deliver results much faster because complex meshing is no longer required. This paper addresses using three different types of force (direct force, torque, and pressure) to calculate values such as Von Mises hotspots and compare the results with mechanical or design phase testing.

Keywords: CT, Computed Tomography, additive manufacturing, X-ray, structural mechanical simulation, FEA, finite element analysis, Von Mises, voxel

1. Introduction

Manufactured and/or structural parts can differ from their ideal geometry due to the manufacturing process. In casting and injection molding, gas or shrinkage pores are possible, while in additive manufacturing you can have sintering or melting. Industrial X-ray computed tomography (CT) allows for the non-destructive detection and visualization of many kinds of defects. With today's CT scanners reaching spatial resolutions down to the micrometer scale, a CT image constitutes an accurate three dimensional digital representation of the scanned part. The spatial distribution and the morphological features of the porosity can thus be assessed in detail [1, 2]. However, the mere presence of porosity does not mean the part is unusable. In fact, even highly porous parts can still be good for a specific purpose. In order to minimize false rejections and thus production costs, it is desirable to assess the "effect of the defect", i.e., to what degree a geometric defect affects the mechanical function. Work has been done to relate the detected porosity to fatigue behavior in aluminum castings [3] and laser-melted parts [4], sometimes combining the pore geometry with heuristic modeling of the pore physics [5, 6] or with analyzing stress concentrations at individual pores using finite element analysis [7]. In this study, we follow a more direct path by feeding the full three-dimensional digital model obtained from the CT scan into a mechanical stress simulation. Earlier work was going in this direction [3, 8]. However, real-life data can exhibit a rather complex geometry (e.g., ragged surfaces or fine-structured porosity), rendering the simulation task technically very challenging. In conventional finite element methods, the simulated domain must be tiled into a mesh of small volume primitives before a simulation can be conducted. This meshing step is crucial for the success of the simulation and can be challenging for complex geometries [3, 8], or even practically impossible [9], since contrary objectives are to be met. First, the mesh must be fine enough to capture all relevant geometric details. Second, the number of mesh cells must not be too large in order to keep the computational effort at a practical level. Third, the mesh cells must conform to certain shape criteria in order to assure the numerical stability of the simulation [10]. Recently, mesh-less and immersed-boundary finite element methods have been used to overcome the meshing problem [9, 11]. Such methods do not require the generation of a boundary-conforming mesh and are therefore suited for the simulation of arbitrarily complex domains as found in porous objects. The use of immersed-boundary methods with CT data [12] for a single pore embedded in the component model is well suited for simulation. Entire highly-porous test specimens (castings) were simulated [13], and results were compared to physical tests. In the present work, 36 specimens with varying porosity were investigated by simulating their complete CT-scanned models on the pore scale using an immersed-boundary method. Simulated is a static load in the linear-elastic approximation followed by the analysis of the von Mises yield criterion. The simulations are matched against quasi-static physical tensile tests. The prediction of fatigue failure is commonly based on the analysis of linear statics with the aid of damage accumulation models (e.g., W¨ohler curves). Thus, it makes sense to examine the static case for a proof of concept, even though more complex failure modes may be more relevant in many practical use cases.

2. Methods

2.1. Test Specimens

In order to assess the effects of porosity, we need an ensemble of specimens with a varying degree of porosity. Modern additive manufacturing technologies allow us to build structures on scales as small as a tenth of a millimeter or less. This enables us to digitally design a 3D model of a specimen, to insert a precisely prescribed porosity structure, and to have it built exactly this way. Two 3D models were used as a basis: a) a standard tensile rod with 5mm diameter and 50mm length, and b) an aeronautic part. The aeronautic part is a bionically optimized structural bracket, originally approx. 150mm long, used in Airbus airplanes, built from Titanium using LaserCUSING_, a laser-melting additive manufacturing process by Concept Laser [14, 15]. The ideal 3D models were digitally deteriorated with different pore configurations. Six different pore configurations were created for each ideal model. The test specimens were built by Concept Laser from an AlSi10Mg aluminum alloy, three copies per pore configuration, resulting in 18 rods and 18 aeronautic parts, i.e., 36 specimens in total.

2.2. Scanning and Geometry Reconstruction

CT scans were taken of the intact parts by the X-ray service provider *diondo* [16], with a resolution of 0.073 mm isotropic voxel size. The software VGSTUDIO MAX by Volume Graphics [17] was used to reconstruct the material surface from the CT scans, including all internal porosity.

2.3. Simulation Static

Structural mechanics simulations were performed based on the reconstructed surfaces from the CT scans. A linear-elastic isotropic material model was used with a Poisson's ratio of 0.3, which is a common choice for metals. For the tensile rods, one end was fixed and the other end was subject to a unit pulling force. For the aeronautic parts, the ground plate was fixed at the screw holes, and a force was applied to the inner surface of the large bearing ring in a direction 29 degrees inclined from the ground plate. The simulations were performed using the structural mechanics module of VGSTUDIO MAX by Volume Graphics [17], which is optimized for working with complex domains like porous material.

2.4. Physical Tensile Tests

Quasi-static tensile tests were conducted on all specimens. The specimens were elongated in uniaxial stress until complete rupture (tensile rods) or until the occurrence of the first crack (aeronautic parts). The bearing force was measured as a function of elongation and recorded into a stress-strain curve. The tensile rods were threaded at their ends and fixed in a standard specimen holder. The aeronautic parts were screwed with their ground plate to a massive base while a pulling force was applied with a bolt fitted through the bearing ring. The base was tilted, resulting in a force direction of 29 degrees from the ground plate (equal to the inclination of the bar structures). Since we are interested in the occurrence of the first crack, the tests of the aeronautic parts were stopped manually as soon as the first cracks were visible. Due to the human reaction time and the lag of the machine control, many specimens were completely ruptured nevertheless.

2.5. Data Analysis

2.5.1. Prediction of Tensile Strength and Crack Location

The static structural mechanical simulations provide the local stress tensor at all points within the specimen. Further analysis is necessary in order to deduce predictions about the real failure behavior. Our aim is to predict both the location of the first crack and the tensile force necessary to initiate the first crack. The analysis is based on the Mises stress _*M*, a scalar quantity, which we calculate at any point in the object from the simulated stress tensor. The Mises stress is a commonly used predictor of yielding in metals. When it exceeds a critical value known as the material's *yield strength*, the material can be assumed to leave the elastic regime and start yielding. All simulations were performed with a unit total applied force. In the tensile tests the force increases monotonically, up to the first damage. Given that the Mises stress scales proportionally to the applied force (which is assumed in the linear-response approximation of the material and for small enough deformations), the yield strength is reached first at the location where the Mises stress at unit force takes on its global maximum. This is where we predict the first crack to initiate: xcrack = argmax x_V _*M*(x), (1) where V is the volume domain of the simulated part. Further, when comparing different specimens, the total force necessary to initiate the first crack, *F*crack, is predicted to be proportional to the inverse of the maximum Mises stress at unit force: *F*crack _ 1 maxx_V _*M*(x) . (2) In practical simulations, we often find a few local stress maxima with stress values only slightly less than the global stress maximum. In the following we will denote local stress maxima by the more intuitive term *hot-spots*. The used software automatically detects all hot-spots and sorts them by their Mises stress. We denote the Mises stress at the first hot-spot by HS1. It is equal to the global Mises stress maximum. The subsequent hot-spots HS2, HS3, etc., are the next largest local maxima in descending order. Due to various uncertainties in modeling, imaging, surface reconstruction, and simulation, the first crack seen in the physical experiment does not always happen at HS1, but it will happen at any of the first few hot-spots. To level out the uncertainties to some amount, we take the average of the largest *N* hot-spots _HS_*N* =*N* _*n*=1 HS*n*, (3) and use this to predict the tensile strength instead of equation (2): *F*crack _ 1 _HS_*N* . (4) We choose *N* = 3 in this study. In this study, no attempt is made to predict the proportionality factor in equations (2) or (4). One may naively assume it could be determined by substituting the (known) yield strength of the material in the equations. In practice, various systematic limitations in both the accuracy of imaging and analysis, and the modeling of the crack physics exist. For the former, cracks will actually initiate at small tips induced by microscopic surface roughness. Even a high-resolution CT scan cannot resolve all relevant micro-structures. The numerical simulation is done with finite spatial resolution, as well. For the latter, the employed linear elastic model does not capture all effects happening at a sharp tip, just before a macroscopic crack occurs. In reality, ductile effects are believed to alleviate high stress concentrations to a certain amount. In this sense, the described method is a heuristic one.

2.5.2. Analysis of Tensile Tests

The tensile tests result in stress-strain curves. The maximum point in a stress-strain curve indicates the ultimate tensile stress, i.e., the point of complete rupture where the specimen cannot bear any force anymore. However, the method described in the previous section is designed to predict the occurrence of the first crack. For the tensile rods, this happens to coincide with the ultimate tensile stress, probably due to an avalanche effect. For the aeronautic parts, due to their more complex geometry, the occurrence of the first crack does not always cause immediate total destruction. Instead, the stress at the first crack is relieved and redistributed to other structural elements, allowing for a further increase in force. These events are visible in the stress-strain curve as sudden drops in stress followed by a recovery. For the aeronautic parts, the maximum in the stress-strain curve before the first sudden drop was determined and identified with Fcrack in equation (4).

3. Results and Discussion

3.1. Tensile Strength

The physically measured fracture force is plotted versus the prediction. The numeric cross-correlation is 0.96 for the tensile rods and 0.78 for the aeronautic parts. Some interesting observations shall be noted: The rod configurations 75(A,B) and 125(A,B) differ in the number of pores (75 vs. 125 pores). Looking at the porosity only, one would assume that the 75(A,B) configurations are generally more stable than 125(A,B). However, this is not the case. First, 125B is actually approximately as stable as 75B, both in physical tests and in the simulations. Second, 125A is even more stable than 75B in the physical tests. This is in accordance with the simulations when regarding the average of each three specimens. With the aeronautic parts, the variations in stability are significant even within specimens of the same pore configuration. The variations due to the manufacturing process seem to be relevant and their effect on stability can be predicted by the proposed method.

3.2. Crack Location

As detailed in section 2.5.1, the hot-spots (local maxima) in the simulated stress distribution, HSi, can be treated as a list of predicted crack locations, sorted by likelihood. The likelihood decreases with increasing hot-spot index i. HS1 is the most likely guess, HS2 is the second likely guess, and so on. For the aeronautic parts, the correlation of the hot-spot locations with the physically observed crack locations was investigated in detail. It turns out that for 12 out of the 18 specimens, a crack did actually occur at the first or second most-likely predicted location. For 3 specimens, cracks did occur within the top-ten most-likely predicted locations. Only for the remaining 3 specimens (incidentally, all 250D specimens), there was no coincidence between any predicted and real crack location.

4. Conclusions and Outlook

We described a method to predict the location of the first crack and the tensile force necessary to initiate the first crack for porous aluminum parts. The method is based on a high-resolution CT scan of a part, a subsequent static structural mechanics simulation of the scanned model, and, finally, the analysis of local maxima of the simulated stress distribution. We showed in validating experiments that the predictions are very accurate, despite systematic limitations in the spatial resolution of the scan and the simulation, and the simplified modeling of the crack physics. In this sense, the method shows to be robust. Only static loading was considered here even though fatigue due to dynamic loading is of greater interest in many realistic scenarios. Since fatigue is commonly modelled on the basis of quasi-static stress distributions with the aid of cumulative damage models, we assume that the good correlation between simulation and experiment will also hold for fatigue due to time-dependent loading. The experiments were conducted with additively manufactured aluminum parts where porosity was deliberately built in. The analysis of variations among specimens within the same pore configuration suggest that the effects of the non-intended porosity occurring in the laser-melting additive manufacturing process can be predicted, as well. Whether this method also works for typical porosity occurring in casting and molding remains to be studied. The cast specimens examined in [13] were too similar in their ultimate tensile strength to see a significant correlation with the simulations, yet a moderate correlation was observed. The similarity of tensile strength in that case was due to the pores being in the same location (centre) of all studied samples. Yet unpublished data from eight die-cast tensile rods, however, suggest that the method can be applied here without changes.

Acknowledgments

We thank Airbus Emerging Technologies & Concepts for the permission to use the bionically optimized bracket model. We thank diondo X-ray Systems and Services, Hattingen, Germany, for performing the CT scanning. We would also like to thank Henry Weber at Volume Graphics for creating the renderings and photographs, and Joel Micah Miller for creating the photographs.

References

[1] E. Maire, P. J. Withers, Quantitative X-ray tomography, International Materials Reviews, 59(1) (2014) 1–43.
[2] A. du Plessis, P. Rossouw, X-ray computed tomography of a titanium aerospace investment casting, Case Studies in Nondestructive Testing and Evaluation 3 (1025) 21–26.
[3] Q. Wan, H. Zhao, and C. Zou, Effect of micro-porosities on fatigue behavior in aluminum die castings by 3D X-ray tomography inspection, ISIJ international 54(3) (2014) 511–515.
[4] S. Siddique, M. Imran, M. Rauer, M. Kaloudis, E.Wycisk, C. Emmelmann, F.Walther, Computed tomography for characterization of fatigue performance of selective laser melted parts, Materials & Design 83 (2015) 661–669.
[5] J.P. Weiler, J.T. Wood, R.J. Klassen, E. Maire, R. Berkmortelc, G. Wangc, Relationship between internal porosity and fracture strength of die-cast magnesium AM60B alloy, Mater. Sci. Eng. A 395 (2005) 315–322.
[6] C.H. Cceres, B.I. Selling, Casting defects and the tensile properties of an Al—Si—Mg alloy, Mater. Sci. Eng. A 220 (1996) 109–116.
[7] G. Nicoletto, R. Konen, S. Fintova, Characterization of microshrinkage casting defects of AlSi alloys by X-ray computed tomog-raphy and metallography, Intern. J. of Fatigue 41 (2012) 39–46.
[8] E. Maire, A. Fazekas, L. Salvo, R. Dendievel, S. Youssef, P. Cloetens, J.M. Letang, X-ray tomography applied to the characteriza-tion of cellular materials. Related finite element modeling problems, Composites sci. and technology 63(16) (2003) 2431–2443.
[9] V.P. Nguyen, T. Rabczuk, S. Bordas, M. Duflot, Meshless methods: a review and computer implementation aspects, Mathematics and computers in simulation 79.3 (2008): 763-813.
[10] J.R. Shewchuk, What Is a Good Linear Finite Element? Interpolation, Conditioning, Anisotropy, and Quality Measures (Preprint), University of California at Berkeley 73 (2002).
[11] D. Schillinger, M. Ruess, The Finite Cell Method: A Review in the Context of Higher-Order Structural Analysis of CAD and Image-Based Geometric Models, Arch. Comp. Meth. Engin. 22(3) (2015) 391–455.
[12] S. Duczek, H. Berger, U. Gabbert, The Finite Pore Method A new approach to evaluate gas pores in cast parts by combining computed tomography and the finite cell method, Int. J. Cast Metals Research 28(4) (2015) 221–228.
[13] A. du Plessis, I. Yadroitsava, S.G. le Roux, I. Yadroitsev, J. Fieres, C. Reinhart, P. Rossouw, Prediction of mechanical performance of Ti6Al4V cast alloy based on microCT-based load simulation, Journal of Alloys and Compounds 724 (2017) 267–274.
[14] Airbus, Inc., Printing the future: Airbus expands its applications of the revolutionary additive layer manufacturing process, web article (2014), http://www.airbus.com/presscentre/pressreleases/press-release-detail/detail/printing-the-future-airbus-expands-its-applications-of-the-revolutionary-additive-layer-manufacturi/ (loaded May 17 2017).
[15] Concept Laser GmbH, Lichtenfels, Germany, https://www.concept-laser.de (loaded May 17 2017).
[16] diondo GmbH, Hattingen, Germany, http://http://www.diondo.com
[17] Volume Graphics GmbH, Heidelberg, Germany, https://www.volumegraphics.com (loaded May 17 2017).

ULTRASONIC PULTRUSION IN-LINE CURE MONITORING

When sound hits a window

CANADIAN INSTITUTE FOR NON-DESTRUCTIVE EVALUATION

NDT in Canada
NDTC 2018
Canada's NDT Conference

June 19 - 21
Halifax Convention Centre
Halifax, Nova Scotia

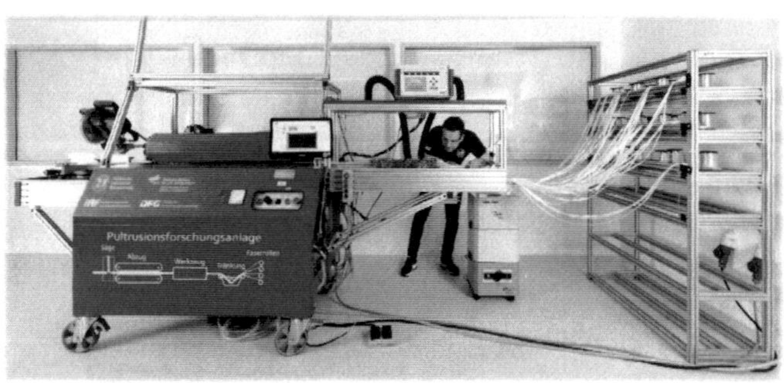

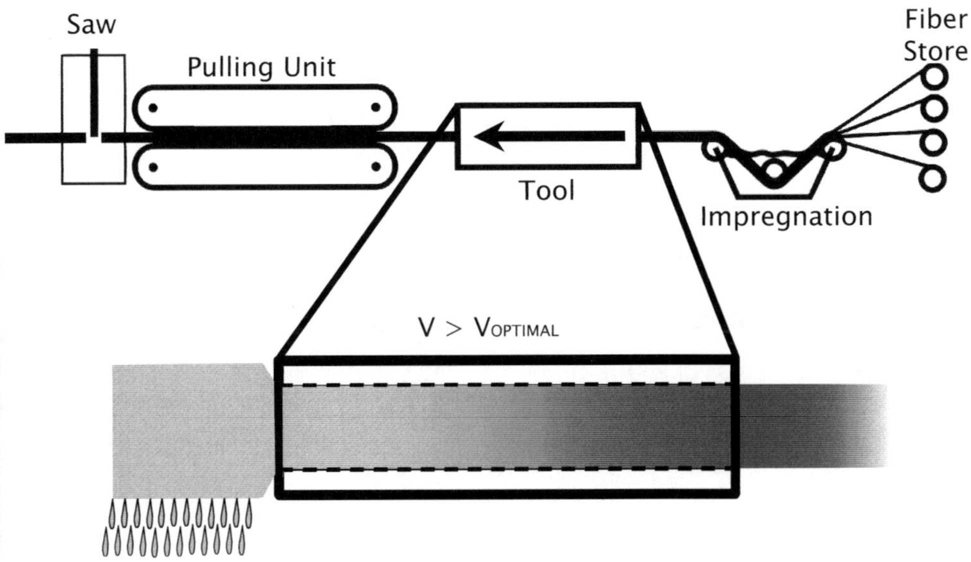

NDT in Canada 2018 | June 19-21 | Halifax, NS

3

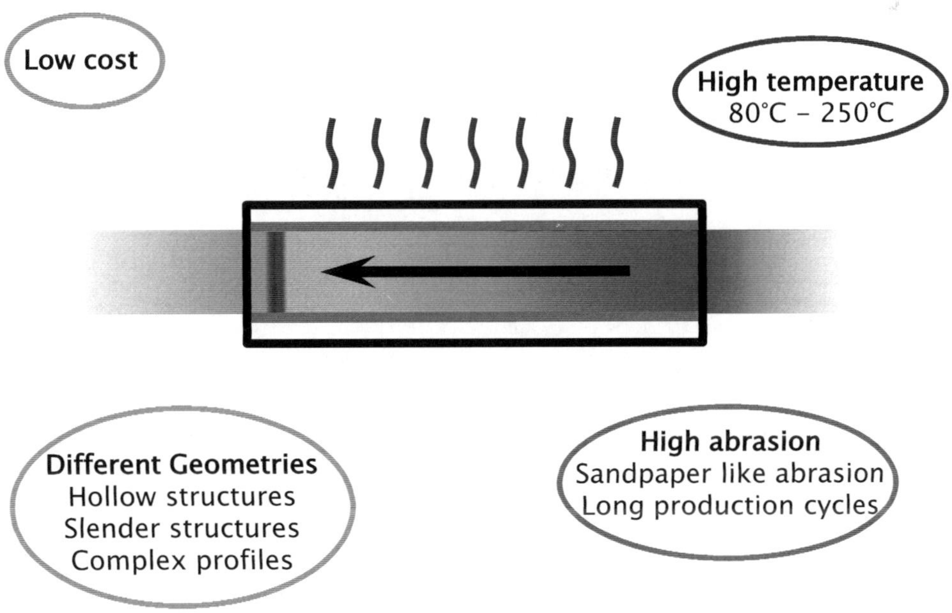

NDT in Canada 2018 | June 19-21 | Halifax, NS

4

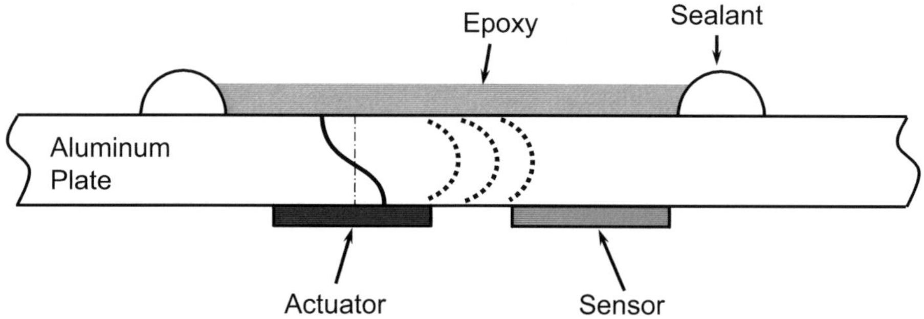

NDT in Canada 2018 | June 19-21 | Halifax, NS

5

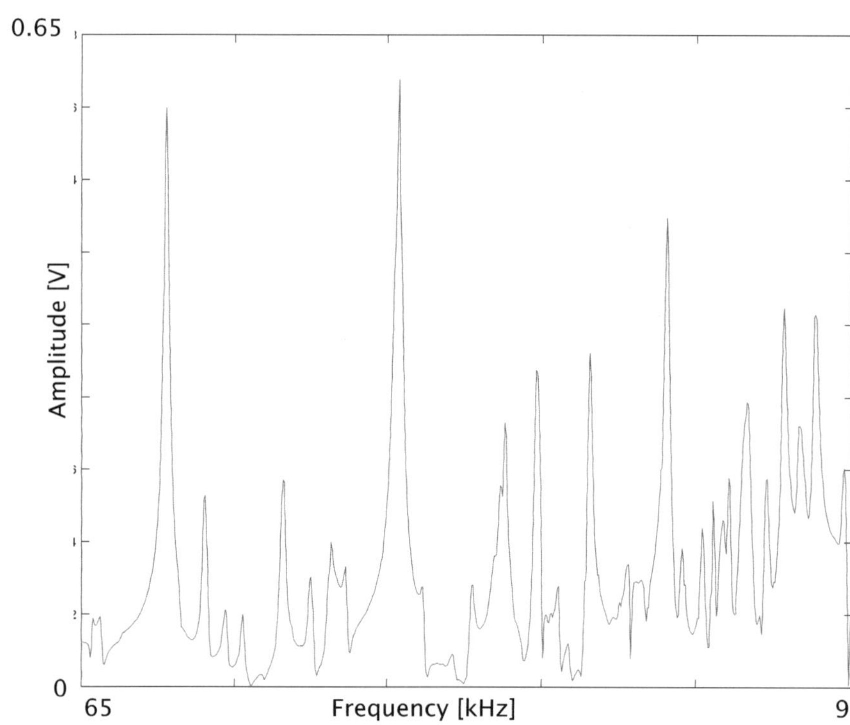

6

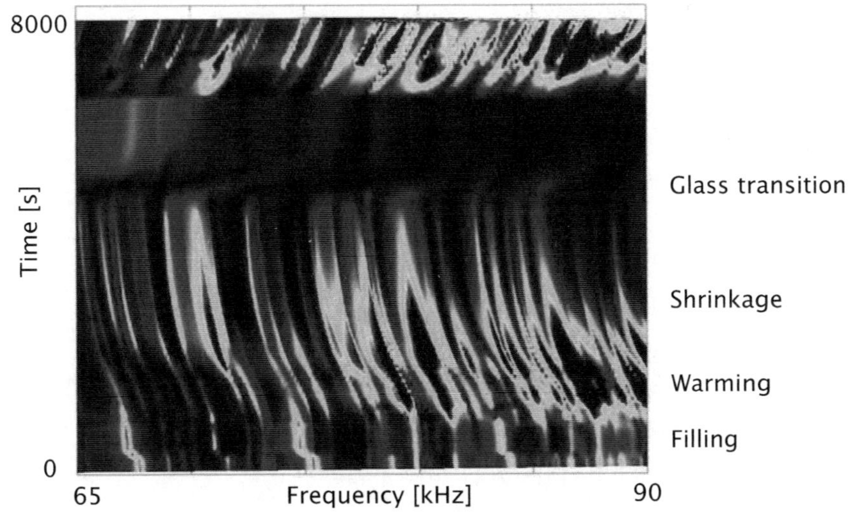
8000
0
Time [s]
65
Frequency [kHz]
90
Glass transition
Shrinkage
Warming
Filling

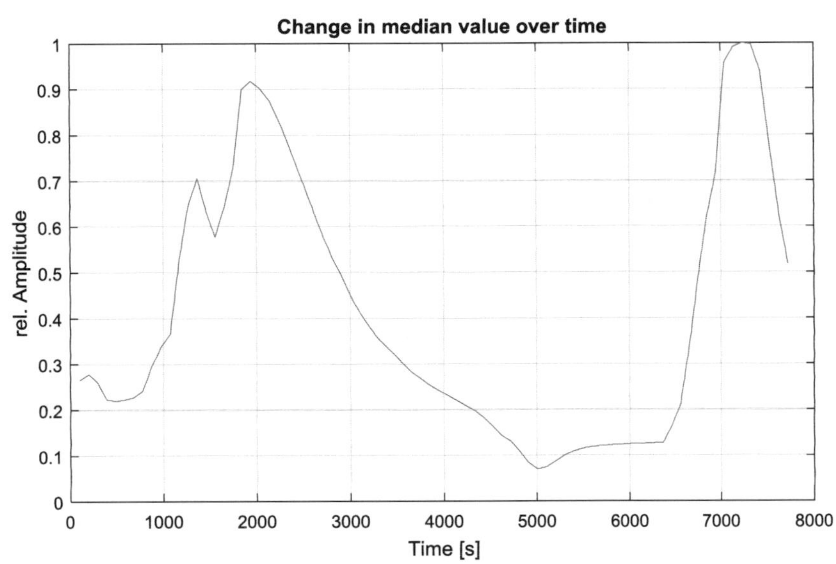
Change in median value over time
rel. Amplitude
Time [s]
0
0.1
0.2
0.3
0.4
0.5
0.6
0.7
0.8
0.9
1
0
1000
2000
3000
4000
5000
6000
7000
8000

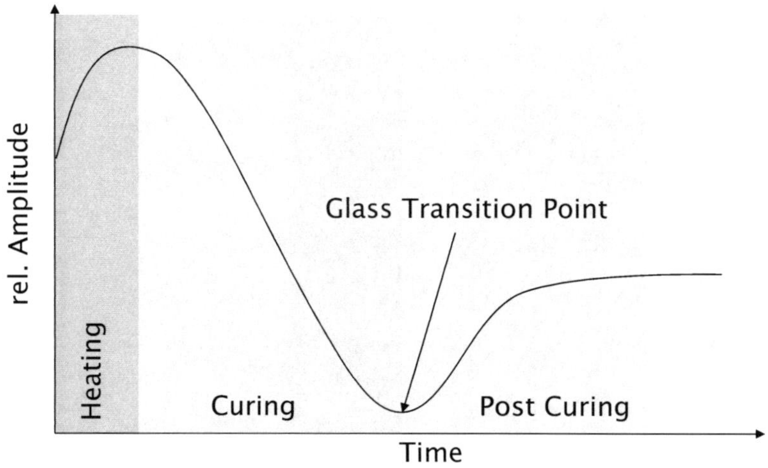

NDT in Canada 2018 | June 19-21 | Halifax, NS

9

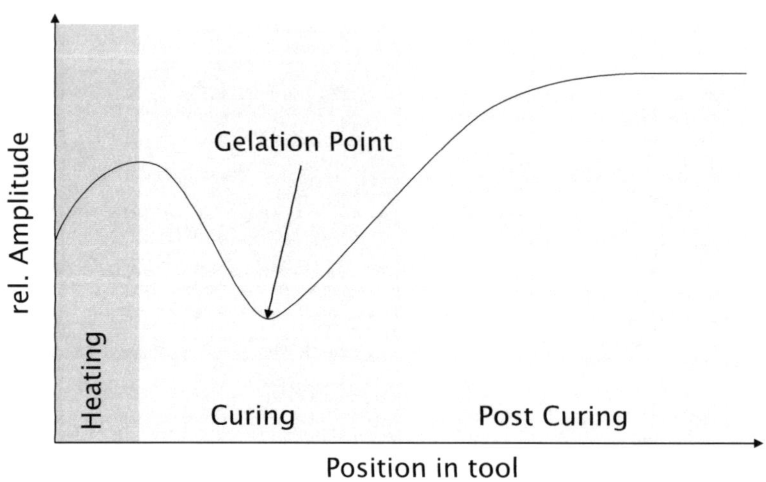

NDT in Canada 2018 | June 19-21 | Halifax, NS

10

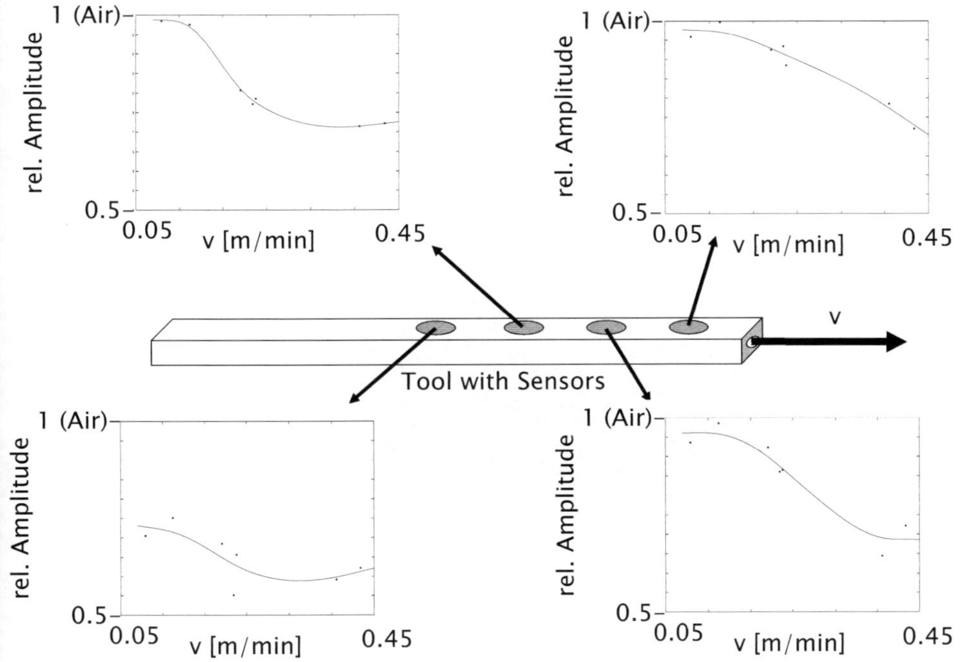

Resonant ultrasonic spectroscopy has a number of advantages against other methods:

- Minimal invasive (by using pressed-in instead of glued sensors)
- No direct contact between tool and sensor (no abrasion)
- Highly temperature resistant
- No high speed digital to analog converters necessary

Future concepts:

- ❖ Further reducing the system size
- ❖ Stabilizing the measurements
- ❖ Optimizing control algorithms

Questions?

?

Thank you for your attention!

NDT in Canada 2018 | June 19-21 | Halifax, NS

15

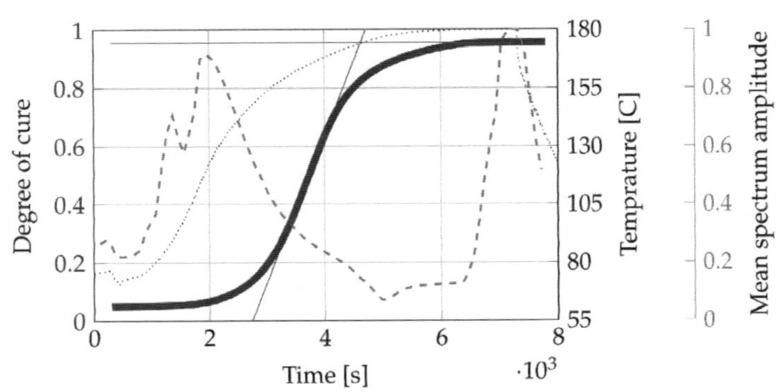

Figure 4. Measured temperature (black-dotted line) and calculated degree of cure (blue line) of the epoxy during the experiment with added support lines (in red) to determine the glass transition point (intersection). The mean spectrum amplitude of Figure 3 is shown in the green-dashed line.

NDT in Canada 2018 | June 19-21 | Halifax, NS

16

Novel Imaging Techniques for Defects Characterisation in Phased Array Inspection

P. Rioux[1], F. Lachance[1] and J. Turcotte[1]
1 Sonatest, Québec, Canada
Phone: +1 418 683 6222, e-mail: sales@sonatest.com

Abstract

Phased Array ultrasonic testing has demonstrated over the years that it has a lot of advantages compared to conventional techniques: fast, reliable, with higher resolution and sensitivity. While it took time to catch up in the field, it is now widely considered as the go-to technique for complex inspections. At the same time, technology has continued to evolve, thus increasing the possibilities with phased array. With increased computing power and better capability to manage large data set of new generation ultrasonic equipment, advanced imaging techniques like TFM are now available. Unfortunately, these techniques are not yet recognized by standards and can't be used alone during critical inspection. On the other hand, they can be a big help in improving defects characterization when used in conjunction with standards-approved methods. In this paper, we will demonstrate that the combination of Phased Array inspection with advanced imaging techniques, can enhance the defects characterization process for common applications of ultrasonic testing.

Keywords: ultrasound testing (UT), phased array (PA), sizing, imaging, linear scanning, sectorial scanning, full matrix capture (FMC), Total Focusing Method (TFM)

1 Introduction

While the phased array ultrasonic equipment has not changed a lot in the past few years, the software advances on signal processing have had breakthroughs. One of the last improvements in the NDT world is the full matrix capture (FMC) and total focusing method (TFM). A new generation of portable devices equipped with these features has followed the industrial systems. The FMC is the ultrasonic acquisition of the area and the TFM applies an algorithm on this acquisition to create an image, where each pixel is a sum of the resulting delays at every coordinate. With this technique, the ultrasonic emission only requires one element, but all elements contribute to the reception. This enhances the focusing capability and the extracted images can show different travel paths and velocities. These transmitted and received patterns are referred to as modes. The NDT community is currently trying to standardise this new approach which has advantages and disadvantages compared to phased array. This paper compares the two techniques and shows how they can become complementary to enhance inspection results where applicable.

2 Basic Principles

2.1 Phased Array

Phased array ultrasonic is based on real A-scans; the amplitude along the propagating beam is predictable according to the laws of physics. The ray tracing area is simple but usually part of a mandatory inspection plan. The UT data recorded will remain the same unless the inspector changes the gain or some other filtering effects. Once a beam inside a phased array scan is generated, this A-scan is defined by the same properties (aperture, frequency, and focal distance) as conventional UT. The criteria of calibration and rejection of this technique are more or less an adaptation of the conventional UT. It is now widespread across the NDT world as the go-to ultrasound technique for complex inspections.

2.2 FMC / TFM

For the full matrix capture (FMC) technique, the raw A-scan data is collected by using a single element as transmitter and all elements as receivers [1]. The TFM imagery is based on coloured pixel representations of a specific area inside the part [2]. This image is based on huge matrices of ultrasonic A-scan obtained through the FMC acquisition process. The TFM result is an image that is not necessarily predictable according to the conventional ultrasonic physics. By itself, it does not have the signal information once generated. However, different image outputs could be extracted from the FMC depending on the chosen analysis mode: longitudinal to longitudinal waves (L-L), longitudinal to transversal waves (L-T), longitudinal to transversal to transversal waves (L-T-T) or simply a suite of transversal waves (T-T and T-T-T). The amplitude and delay of the A-scans are summed, on an X-Y coordinate (with a given resolution), and the colour signature indicates a "presence" regardless of its direction or amplitude. Also, the extracted image would be focused in every point, and because of that, the sizing accuracy should be better compared to a single focusing point, while using the same phased array transducer. While calibration and rejection criteria are not yet defined by standards for these techniques, being able to stock all the FMC A-scans during the acquisition should allow the TFM image to be analyzed on different modes. For example, one of the main advantages of stocking this raw FMC information is that a TFM image can be altered afterwards using different processing tools while the FMC information remains unchanged.

The first section of the paper shows the results comparing a standard linear scan at a 0-degree longitudinal wave (L-wave) with an FMC where L-L TFM images are computed. The second section of the paper shows the results comparing a standard sectorial scan transversal wave (S-wave) with an FMC where T-T-T-T TFM images are computed.

3 Results – Phased Array Linear Scan & TFM

For this section, the phased array linear scan data has been recorded at the same time as the FMC data using the multiscan function of the Sonatest Veo+ instrument. All the phased array linear and TFM imaging data analysis have been performed using Sonatest UTstudio+ software (TFM images are computed over 32x32 and 64x64 FMC apertures).

3.1 Instrument Sensitivity

To begin with, Figure 1 shows amplitude measurements of A-scans extracted from a linear pulse echo (PE) phased array setup (L-scan) and TFM images. The extracted A-scans from the phased array setup have been generated using active apertures of 1, 8, and 16 elements. Six reference holes have been used for the measurement. Diameters are ranging from 2.38 mm SDH at 80% FSH to 0.4 mm SDH. For every acquisition, the signal from the reflector was brought to 80% FSH. For the FMC acquisition, the TFM image is from an L-L mode algorithm, and the amplitude value was taken from the binary digital pixel value, which was converted into the full digital scale of the 16-bit ADC (analogue digital converter) of the equipment and then normalised to 80% FSH.

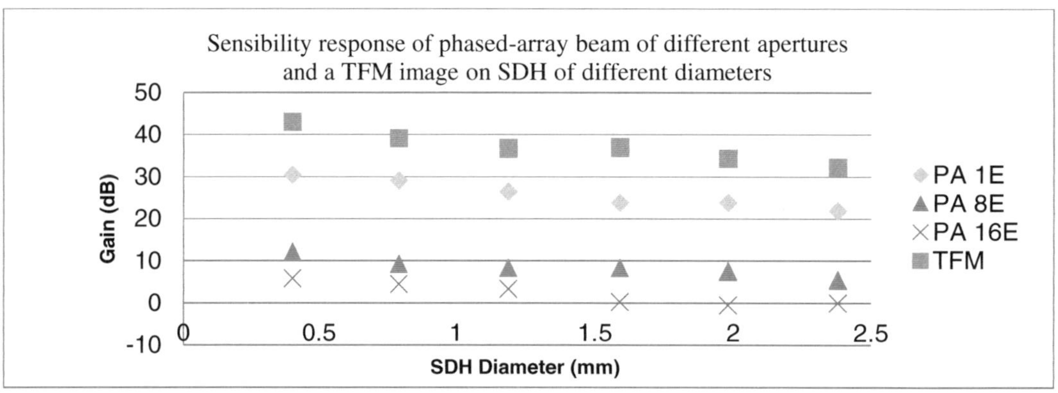

Figure 1: Necessary gain on the ASTM block to get SDH from 0.4 mm to 2.4 mm at 80% FSH.

Figure 1 illustrates four necessary gains in dB: three PE phased array data sets (1, 8, and 16 active elements) and a TFM image, to bring the corresponding SDH to 80% FSH. As one could have expected, the TFM image requires higher gain to bring the defects to the desired amplitude. Figure 2 shows results over the 0.4mm.

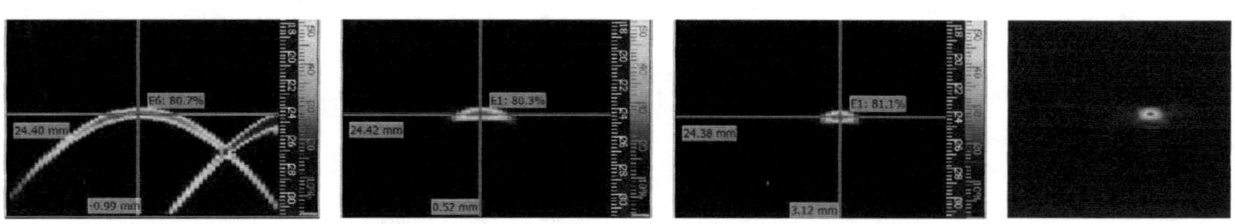

1E PA @ 31 dB | 8E PA @ 12.1 dB | 16E PA @ 5.9 dB | TFM @ 43 dB*

Figure 2: Sensitivity of a 0.4mm SDH, according to the compared techniques
*Reference to the sampled image peak value then converted into FSH and dB.

Generally speaking, for the phased array images, the more elements used for the acquisition, the less amplification is needed; as the laws of physics would predict. The principal advantage of using an aperture of many elements with phased array is the noise reduction; there is an average improvement of 24 dB using an aperture of 16 instead of 1 elements. This is mainly because the more elements are used, the more energy is transmitted into the part for one acquisition.

For the TFM image on the right of Figure 2, the FMC acquisition was produced with the same gain as the 1 element phased array scan (31 dB) and an equivalent gain of 43 dB was found. The colour palette of the TFM image is set automatically using an equivalent palette as standard phased array. Interestingly, even if the equivalent gain used for the TFM image is higher (43 dB) compare to PA (31 dB), the signal-to-noise ratio of the TFM image is much better than the 1 element phased array scan. This signal-to-noise ratio enhancement can also be observed besides the SDH compared to other images (8E and 16E phased array) as seen in Figure 2. The total focusing effects of TFM enhance the signal and build the image by positive interference. Since the UT noise is everywhere and random, its contribution is destroyed by the algorithm and significantly reduced in the TFM images.

Another advantage, derived from the positive interference technique applied in the TFM solution, is that instrument gain applied during the FMC acquisition is far less critical for sizing a defect when compared to phased array. While computing the TFM images over a defective zone, the maximum will be automatically normalised to 100% FSH. Then the user can compensate using software gain with minimal impact on induced noise, regardless of the original instrument gain sensitivity setting. These low noise and high-resolution TFM images are possible because the selected instrument, the Sonatest veo+, records over 16-bit and has a very low noise chain of acquisition. Considering that FMC only pulses 1 element (low energy source) at the time, this weak point is attenuated by a low noise acquisition chain and by the software TFM summation algorithm. On Figure 3, TFM images of the same SDH have been computed from FMC acquisitions done at three different gain levels to show this behaviour, i.e. that over a 60 dB gain span for FMC acquisition, the same 0.4 SDH can be characterised using TFM.

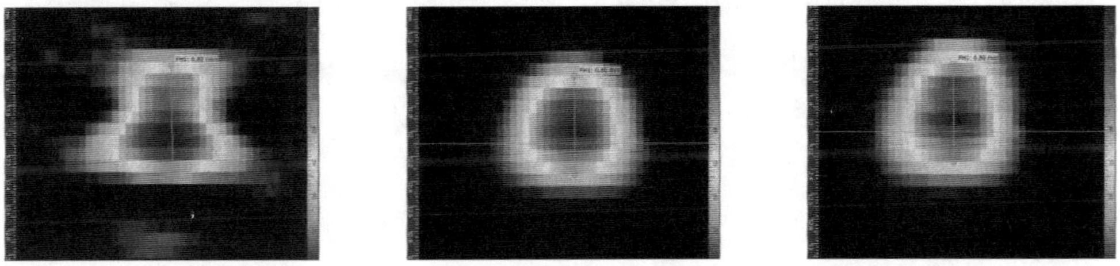

TFM with FMC @ 0dB   TFM with FMC @ 30dB   TFM with FMC @ 60 dB

Figure 3: Sensitivity of a 0.4mm SDH, according to the FMC gain sensitivity

3.2 Sizing Capabilities

Because of the total focusing advantage, the TFM shows an advantage that can be used to help size critical indications. The AWS resolution block has been selected for the next demonstration because it has three SDHs close to one another. The diameter of the SDH is 1.5 mm, and the distance between each of them is 4 mm. Figure 4 compares the linear scan phased array and TFM techniques from the same probe position and SDHs:

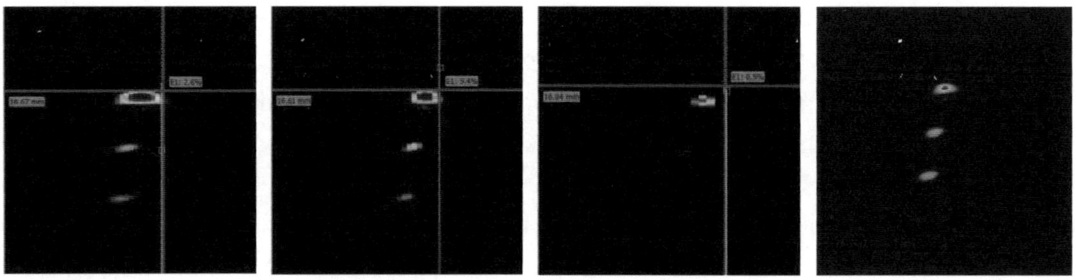

8E. L-scan 16E. L-scan 32E. L-scan TFM image

Figure 4: AWS resolution block with three SDH vertically aligned

For the linear scan, the focal distance was set on the first SDH, and the results show the presence of the three SDHs but the first hole is getting smaller and the beam divergence after the focal point increases as the number of active elements increases. Indeed, the 32 elements aperture (0.8 mm pitch) configuration, hardly shows the last SDH because of that phenomenon but the SDHs can be detected via the other more common linear scanning configurations (16E and 8E). From those results, it is however observed that properties like the number of elements per focal law and the focusing depth, if not properly understood by the user, can affect sizing capabilities of the second and third SDHs. With the FMC acquisition and subsequent TFM imaging, one can successfully size all three SDHs with more precision, even if the amplitude is also lower on second and third SDHs (there is no TCG option in TFM imaging). It also shows that this technique is again less dependant on the scan or probe properties.

3.3 Probe Resolution Parameters

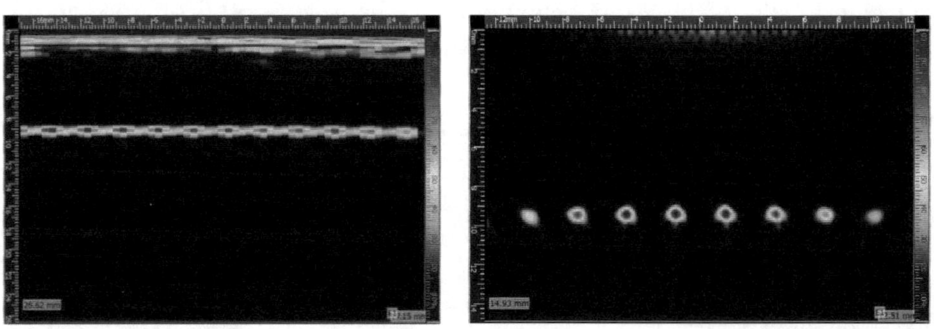

(a) Linear scan (b) TFM image

Figure 5: Lateral resolution

The spatial resolution in the probe axis (lateral resolution) can be considerably improved by recording the FMC and subsequently computing TFM images as well. Figure 5 presents results of both techniques on a standard test block having diameter 1 mm SDHs with 3 mm between each of them. In Figure 5a, we can see that the linear scan will detect the presence of these SDHs properly, but will hardly distinguish the size of these individual SDHs because the amplitude drop between each SDH is less than 6 dB (A-scan amplitude is comprised between 50% and 80% FSH in the holes region). In the case of the FMC acquisition and TFM imaging, the algorithm triangulates the position of all beams which has the effect of improving the lateral resolution limitation from the physical size of the probe elements to the wavelength of the ultrasound beam. We can see in Figure 5b that the TFM image can size the SDHs properly, with a clear amplitude drop between each of them. The focusing ability of this imaging technique proves again that when required, it can become a real improvement to a standard linear scanning solution where resolution is critical.

3.4 Signal Processing Resolution Parameters

The resolution chosen in the TFM algorithm is a critical parameter which can create errors on the output images. Starting from a matrix that has 1024 A-scans (32x32 FMC), you can extract an image of 256 by 256 pixels. Using a resolution of 0.1 mm^2 represents a region of 25 x 25 mm. Increasing to a 4096 A-scans matrix (64x64 FMC) will simply increase the region of interest, not the resolution. On the other hand, decreasing the pixel resolution does not change the centre position, but the software gain needs to be increased to achieve the same result. We can see in Figure 6 that both detection and sizing are not affected if the resolution is increased from 0.1 mm^2 to 0.2 mm^2. The wavelength in this setup being 1.2 mm, a 0.3mm^2 resolution could be used without affecting too much the amplitude. As predicted by the Nyquist theorem you may statistically loose up to 3 dB 50% of the cases. However, as presented in Figure 6, going over that threshold of 0.3mm^2 resolution affects the TFM sizing capabilities. Having access to such resolution software tools is important for precise TFM image analysis.

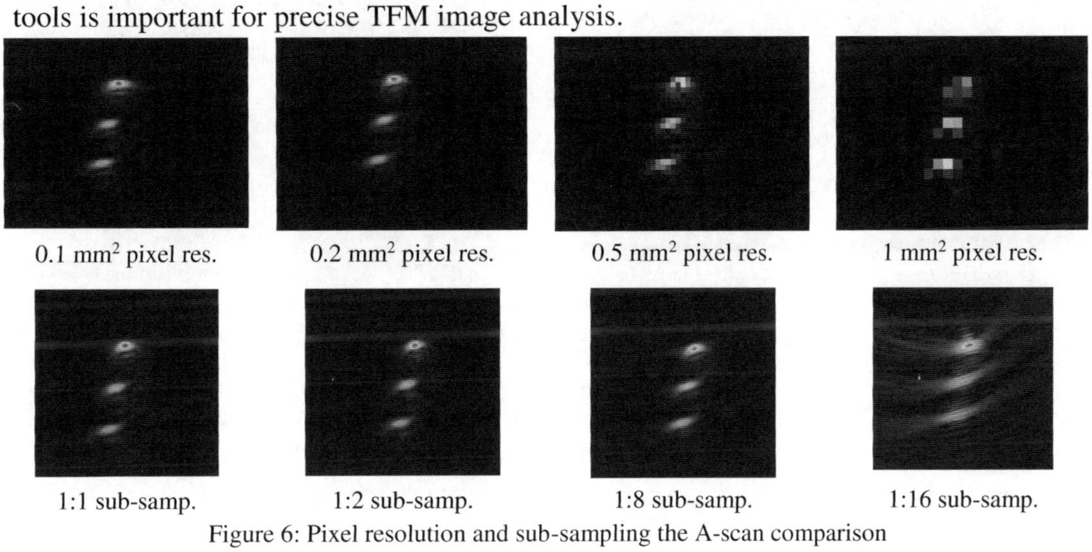

0.1 mm^2 pixel res. 0.2 mm^2 pixel res. 0.5 mm^2 pixel res. 1 mm^2 pixel res.

1:1 sub-samp. 1:2 sub-samp. 1:8 sub-samp. 1:16 sub-samp.

Figure 6: Pixel resolution and sub-sampling the A-scan comparison

A similar comparison can be made by removing samples directly from the raw A-scans. In the next example, the size of a frame could be cut in half to save memory space, without affecting the result. We can see in Figure 6 that the amplitude is not so different from the first subsampling 1:1 compared to the 1:2 but the image is considerably altered for the 1:16 subsampling. In numbers here, using a subsampling of 1:2 for a 50.8mm range will reduce the sample density from 42 samples/mm to 21 samples/mm. It still respects the 5 to 1 ratio of the acquisition frequency to the probe frequency. This sampling rate has virtually no impact at this wavelength of 1.2 mm. The FMC frame size is smaller by 2 MB (4 to 2 MB). For a 300mm long encoded scan, this represents 600 MB of memory instead of 1.2 GB, which is not negligible.

3.5 Encoded Real Crack Sizing Comparison

We can now compare a fairly optimised linear scan to an FMC scan on a real defect. Both techniques use an encoder with a 1mm step resolution over a crack that is 100 mm long. In Figures 7, the linear scan uses an 8 elements aperture, and TFM images are computed in L-L mode algorithm. As expected, both techniques detect the crack and can fairly characterise it.

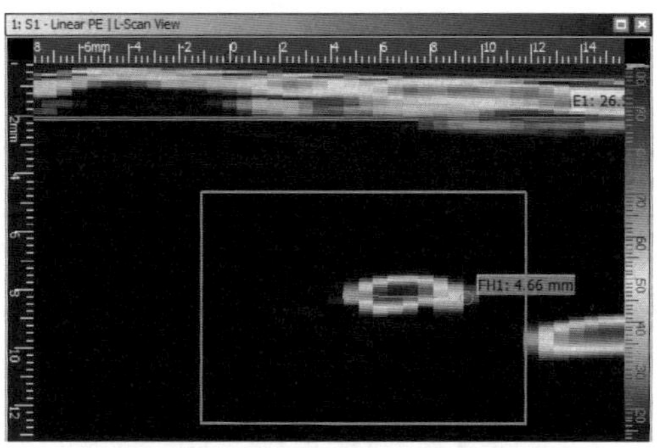

(a) Linear scan - 4.66 mm long

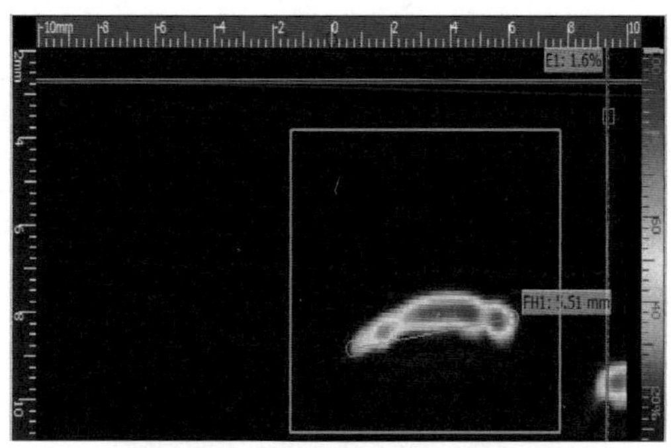

(c) TFM L-L image - 5.51 mm long

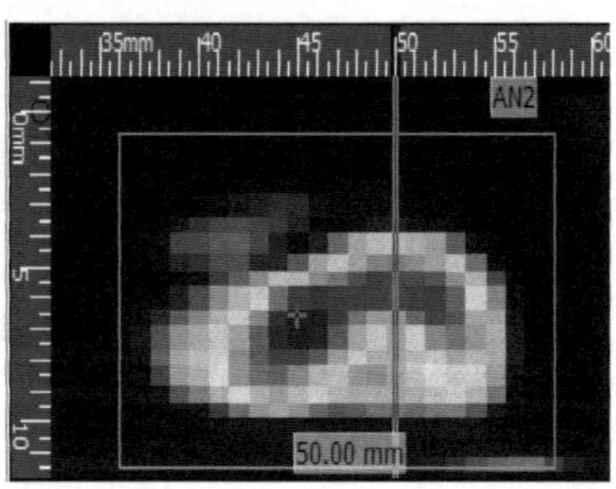

(b) Linear scan – encoded top view

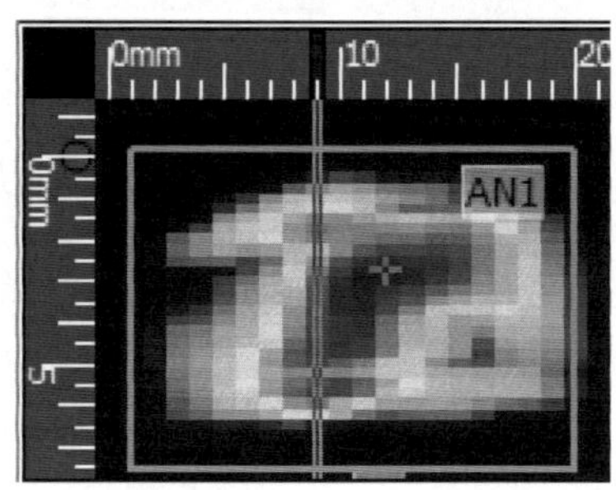

(d) TFM L-L image – encoded top view

Figure 7: Linear scan (8E) and TFM L-L imaging comparison

However, the convex shape of the defect is not showing up in the linear scan view and, consequently, seems to appear smaller in Figure 7 (a) and (b). In Figure 7 (c) and (d), the left portion of the crack is better represented as the crack orientation respond at different angles which are caught by the TFM algorithm. The analysis through time is a critical action where the growth rate is a major threshold parameter for repair. The absolute measurements, such as flaw length, are almost the same since the passive aperture is identical but the projected top view does not render the same dynamic profile. The defect zone measurement at the -6 dB amplitude is 33 mm^2 for the TFM and 37 mm^2 for the linear scan.

4 Results – Phased Array Sectorial Scan & TFM

For this section, the phased array sectorial scan data has been recorded at the same time as the FMC data using the multiscan function of the Sonatest Veo+ instrument. All the phased array linear and TFM imaging data analysis have been performed using Sonatest UTstudio+ software (TFM images are computed over 32x32 and 64x64 FMC apertures).

Since conclusions of the sections 3.1 to 3.4 similarly apply to a sectorial scan comparison, the next section will focus on presenting the results of real weld defects characterisation.

4.1 Real Weld Porosity and Crack Sizing Comparison

Two defects have been scanned using a 55° shear wave wedge and a phased array sectorial scan of 32 elements apertures. Simultaneously, the instrument was recording a 64x64 FMC data set. The Figure 8 presents resulting sectorial scan and high-resolution 0.1mm^2 TFM images. Images have been rearranged for a working comparison.

(a) PA Sectorial Scan
Porosity depth at 4.13mm and 20.3mm^2

(b) TFM TTTT image
Porosity depth at 4.17mm and 20.1mm^2

(c) PA Sectorial scan
Crown Crack depth at 10mm

(d) TFM TTTT image
Crown Crack depth at 8mm

Figure 8: Sectorial scan (32E) and TFM imaging comparisons

Again, it can be seen that the sectorial scan caught both defects accurately as per the depth and sizing because scan has been done using high energy and precise 32 element aperture focused in the defect area. However, as expected the TFM imaging technique shows some improvements in the sizing capability. TFM is especially good to show the shape of the defects. The signal-to-noise ratio is also a slightly better on the TFM images and this is mostly because of the 32 to 64-element aperture.

It is interesting to explain that multiple trials with different TFM algorithms have been tested before generating these images. The LL, LT, TT, LLL, LLT, LTL, LTT, TTT, LLLL and TTTT modes have all been tested using the raw FMC data set in the Sonatest UTstudio+ analysis software, and typically the V-Weld shape in TTTT showed the best results. For weld inspection, when using transversal waves, mode conversion is an important factor to consider. Because of the different defect orientation and geometry characteristics, the TFM algorithms will generate different results. Longitudinal wave inspection mode is less dependant on this behaviour, and the TMF L-L algorithm is mainly used for this one. Because the instrument was simultaneously recording the full FMC raw data set during the recording, data is available for further analysis anytime. This approach is interesting not only for improving sizing capabilities but also to ensure traceability and monitoring defect propagation over time since a data can be re-open any time after the acquisition.

5 Conclusion

Fast, reliable and with higher resolution and sensitivity; ultrasonic phased array technique offers many advantages compared to conventional ones. Also, being now widely accepted in codes and standards phased array remains a very powerful ultrasonic NDT solution to detect and size defects in many different applications efficiently.

This paper has shown that the TFM imaging technique can bring valuable sizing benefits to complement or assist phased array inspection technique. For example, an improved signal-to-noise ratio can become important to discriminate small indications from the noise, and better geometry details can help understanding the potential behaviour of the defect. Sometime, to achieve similar results using phased array technique, apertures of 32 elements with a focus on the defective zone needs to be implemented. Such technique proved to work on many occasions, but it also puts more pressure on the technician to correctly set the acoustic parameters of his instrument. On the other hand, it has been demonstrated in this paper that recording FMC data presents an inferior risk of error than phased array; moreover, it opens the door to a multitude of TFM analysis options.

In conclusion, the approach proposed in this paper supports the methodology of using ultrasonic phased array technique for detection and sizing of defects as it remains today a very efficient and code compliant NDT inspection technique. Using FMC recording and TFM imaging as a detection and sizing technique would also be compliant, but would generate a large amount of data and this data management is still today a concern in the industry. Since the phased array technology proved to cover detection, sizing and traceability requirements, performing FMC/TFM technique only if required over critical defective zones represents a much more efficient solution. According to this approach, an NDT technician can scan with confidence using an approved phased array technique that generates manageable data sets all day long. Then, if inspector feels challenged by the presence of a defect in a critical zone of his job task, high resolution (down to $0.1mm^2$) TFM imaging and analysis tools are subject to help making that call with confidence. That being said, it has been observed in this paper that the TFM image processing must respect a minimum resolution threshold in order to avoid potential sizing errors. Taking this observation into account, performing a TFM analysis over a set of raw FMC recorded data will ensure the inspection decision can be tracked back for future assessment, as it is required by the traceability concept.

Since the usage of this novel FMC/TFM approach gets more and more mainstream, standard procedures are being reviewed by different NDT committees to help NDT technicians work with this new technology. Finally, it is clear that the trend for innovative NDT ultrasonic manufacturer is to invest in this technology; this shall also contribute to making this technology more accessible to NDT professionals.

References

[1] C. Holmes, B.W. Drinkwater, P.D. Wilcox , "Post-processing of the full matrix of ultrasonic transmit-receive array data for non-destructive evaluation," NDT & E International, pp. 38:701-11, 2005.

[2] F. Reverdy, G. Benoist, L. Le Ber, "Advantages and Complementarity of Phased-Array Technology and Total Focusing Method," in 19th World Conference on Non-Destructive Testing, Munich, 2016.

Calibration Procedure Developed for FMC-TFM Based Imaging System

Michael Wright, Jonathan Lesage, Mohammad Marvasti, Steven Peters, Mike Matheson

Acuren Research and Application Development Group, Cambridge, Ontario, Canada

Abstract

In recent years, there has been an increasing interest in the Full Matrix Capture (FMC) data acquisition method within the Non-Destructive Testing (NDT) industry. The FMC approach involves transmitting with each element in a phased array probe in sequence, while capturing the response with all elements. The captured data can be post-processed to produce detailed, high resolution images of the test piece, most commonly by applying the Total Focusing Method (TFM), which effectively focuses ultrasonic energy at every pixel. The resulting images are typically much easier to interpret than traditional phased array sweeps and can lead to a higher probability of detection of indications while decreasing data analysis time. It seems likely that FMC-TFM based systems will eventually come to dominate the NDT industry, provided that a proper calibration procedure can be developed to meet existing inspection codes. To this end, a novel, measurement based, calibration procedure developed by Acuren's Research and Application development group is presented, implemented and assessed based on phased array ultrasonic code requirements. The results of the present study indicate that the proposed calibration procedure can effectively calibrate amplitudes close to the center of the aperture, though cannot meet existing code requirements over the entire aperture for shallow targets. The source of the incomplete amplitude compensation is identified and further research to mitigate these effects are proposed.

1. FMC – TFM Introduction

FMC data acquisition system makes full use of multi-element transducers in a pulse-echo mode. Each array element is used in both transmission and reception to generate an array of raw A-scans from every transmitter-receiver pair. This is achieved by stepping though elements for transmission, one at a time, and receiving on all elements. This transmission/reception sequence is illustrated in Figure 1 for a 4-element transducer which creates a 4×4 matrix of time-domain A-scans.

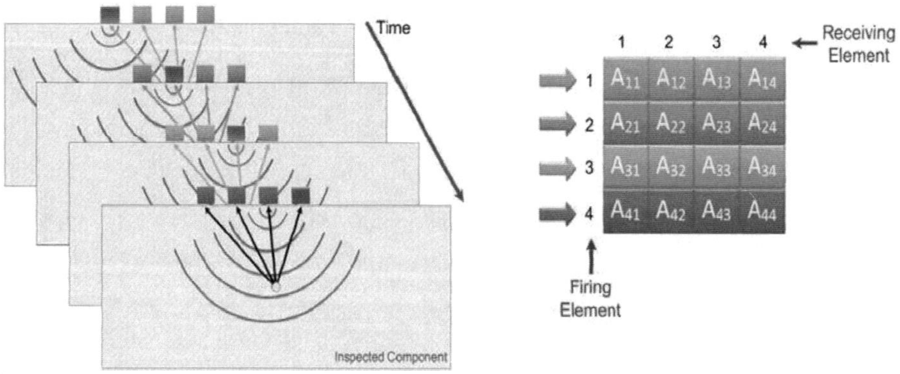

Figure 1 FMC Acquisition Using a 4-Elemenet Probe.

The acquired A-scans are then synthetically focused at every point in the test piece to produce a high quality image using the Total Focusing Method. The TFM algorithm builds an image by first calculating time it takes for an ultrasonic wave to travel between each transmit/receive element pair to each pixel point in a domain and then assigning the amplitude corresponding to the calculated travel time from the A-scan to the corresponding pixel point. Figure 2 illustrates the described imaging process for two pixel points (p and q) in a domain using A-scan recorded from 1-N element pair.

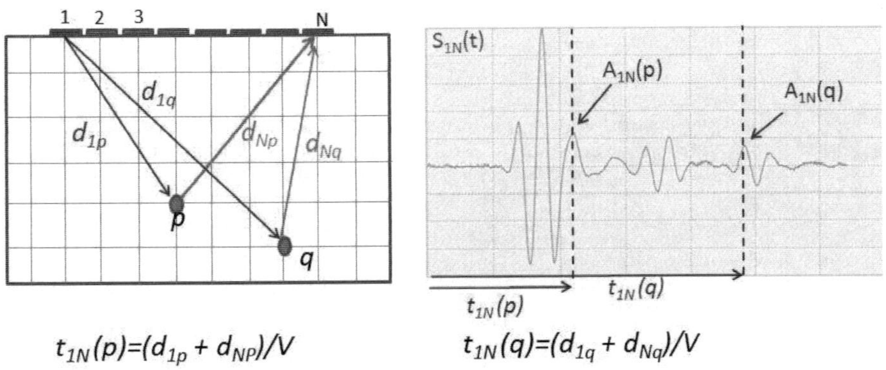

Figure 2 Illustration of the Total Focusing Method.

Once all of the possible image frames are rendered using all transmitting-receiving pairs (N×N frames), they are stacked together to create the final image. The image intensity using the stacking process for any pixel point in the domain (with x and y coordinates) can be presented as shown in equation 1:

$$I(x,y) = \sum_{m=1}^{N}\sum_{n=1}^{N} A_{mn}\big((d_{m,(x,y)} + d_{n,(x,y)})/V\big) \qquad \text{Eq. 1}$$

Using the above stacking routine, high amplitude values get stacked at locations where reflectors are present while very low amplitudes get stacked at locations where no reflectors exist. This procedure results in a very sharp, high quality image, owing to the fact that all of the collected A-scans are effectively focused at every pixel point in the domain. Figure 3 shows the difference between a TFM images built using a single Frame (1 Transmitting Element, N receiving Elements) to a TFM image built by stacking N frames (N Transmitting Elements, N receiving Elements).

Figure 3 shows that, although a sharp, high quality image has been obtained by application of the TFM, image intensities are not uniform across the imaging range. The four indications originate from holes which were the same size and differing only in terms of their relative position with respect to the center of array aperture. This can be largely explained by the effects of element directivity and beam spread. Ultrasonic energy decreases away from the center of a transmitting element – this effect is referred to as element directivity. As a consequence of directivity, reflectors are insonified with decreasing levels of energy, the farther they are from an element's central axis (as measured by angle with respect to the element's centerline). In addition to the effects of directivity, ultrasonic energy decreases with distance from the element's surface. The loss of energy with increased travel path is due to the combined effects beam spread – caused by the wave front spreading out to cover larger volumes as the wave progresses, and material attenuation due to scattering, and dissipative sources. In steels and most other commonly inspected materials, the loss of energy with travel path is dominated by the beam spread effect. Both directivity and Beam Spread lead to the distribution of ultrasonic energy, generated by an element to be non-uniform over the inspection region.

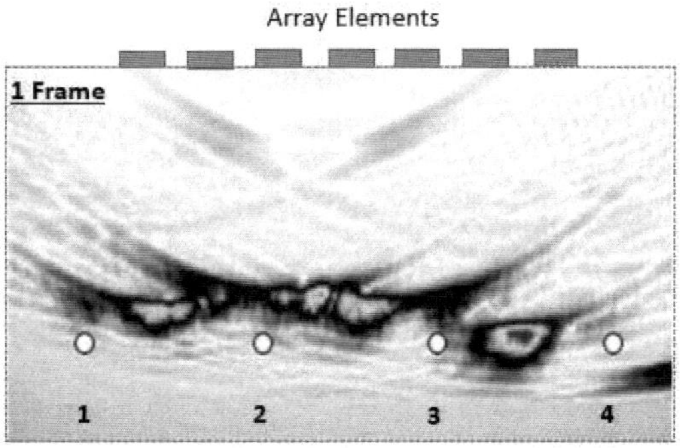

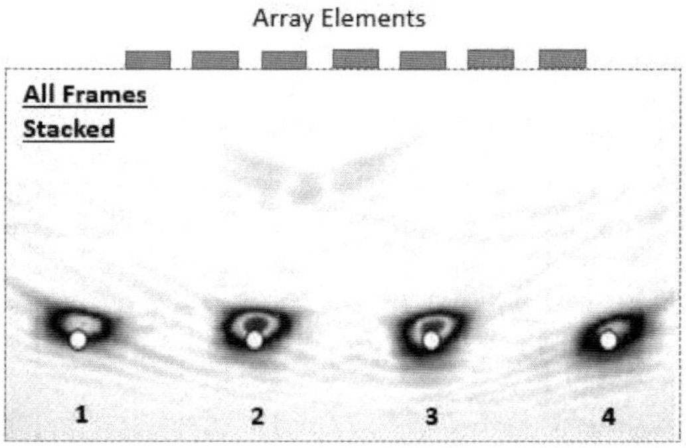

Figure 3 Effect of Stacking TFM Frames.

It can be seen in Figure 3 that the holes near the center of the aperture are covered by all elements with relatively small angles with respect to each element's centerline – this causes the middle holes to be struck with higher energy from each element as compared to the holes located at the edge of the aperture which are covered by larger angles (on average) from all transmitting elements. Figure 4 shows the sound path difference from the last element in the array to the first and second holes. It is readily observed that the path from the last element to the first hole, is longer and at a larger angle with respect to the element's centerline than the path to the second. Consequently, the last element's beam profile strikes the second hole with greater amplitude than the first hole, leading to a higher amplitude in the rendered image.

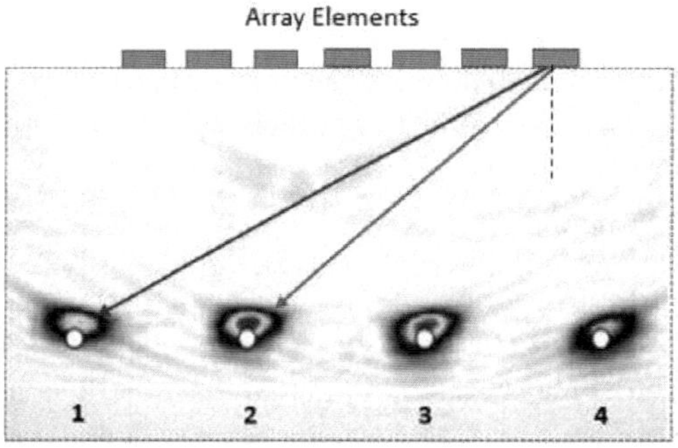

Figure 4 Directivity - Beam Spread Effects on TFM Image Intensity.

A calibration process is required in order to compensate for directivity and beam spread effects at all locations within the imaging/inspection region so that the image intensity (amplitude) obtained from a reflector is solely dependent on the size and shape of the reflector and not its spatial location with respect to the center of aperture. In other words, an appropriate amplitude calibration procedure would cause indications from holes to have the same image intensity regardless of their location within the image.

2. Element Directivity – Beam Spread Quantification

Efforts to correct TFM images based on theoretically derived, approximate formula have been presented in the literature [1], [2], [3]. These formula approximate array elements as point sources which generate ultrasonic pulses containing a single frequency (taken as the center frequency). Beam profiles computed using these approximate relations can be inaccurate due, primarily to the following: 1) in reality the array elements are not point sources and their shape and size in both the active and passive directions is ignored and 2) the wave generated by the elements contains a band of frequencies. In addition, application of the above referenced formulae cannot possibly account for deviations of element properties from their nominal values - most significant of which are: the element width, element spectrum and uniform displacement over the element's surface. For the above stated reasons, it is preferable to directly measure the beam profile, where possible.

A new approach is introduced in the present study to directly estimate the average beam profile (taken to be representative of all elements in the array once element sensitivity is calibrated for) using measured A-Scans obtained from the specular reflection between all element pairs. The theoretical basis of the new approach is explained in this section, testing the approach in terms of a calibration process and then correction of the TFM images to use the calculated directivity and beam spread effects are presented in the next sections.

Consider a multi element array in contact with a block with a known thickness *Th*, as shown in Figure 5. The beam path from the center of the first element to the center of the '*n-th*' element of the aperture after being reflected from the back wall is shown using two blue arrows. Beam angles and the total sound path length between the two elements can be obtained using Equations 2 and 3 respectively where *p* is the elements pitch.

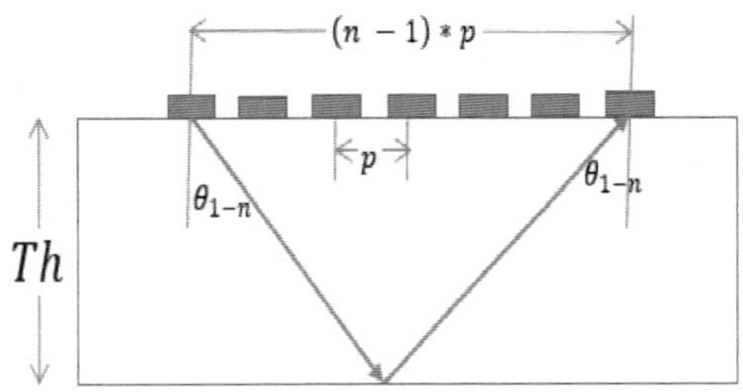

Figure 5 Array Elements Beam Paths to the Backwall.

$$\theta_{1-n} = \tan^{-1}\frac{(n-1)p/2}{Th} \qquad \text{Eq. 2}$$

$$d_{1-n} = \frac{2 \times Th}{\cos\theta_{1-n}} \qquad \text{Eq. 3}$$

A-Scan signal recorded by pulsing on element *1* and receiving on element *n* (S_{1-n}) contains a peak amplitude at time $t = d_{1-n}/V$, equivalent to the time it takes for the sound to travel along d_{1-n} path.

The peak amplitude can be modeled mathematically using Equation 4 where A_1 and A_n represent elements sensitivity, *R*, represent angle dependent reflection coefficient and *C* represents the beam profile as a result of sound wave propagation off the center of the element (directivity) and travel path (beam spread).

$$A_{1-n} = A_1 \times C(\theta_{1-n}, d_{1-n}) \times R(\theta_{1-n}) \times C(\theta_{1-n}, d_{1-n}) \times A_n \qquad \text{Eq. 4}$$

The above equation can be rearranged to obtain the beam profile for Θ_{1-n} angle and d_{1-n} distance combinations:

$$C(\theta_{1-n}, d_{1-n}) = \sqrt{\frac{A_{1-n}}{A_1 A_n R(\theta_{1-n})}} \qquad \text{Eq. 5}$$

The beam profile can be calculated using Equation 5 by knowing the right side of the equation: 1) A_{1-n} can be measured as described above; 2) Reflection coefficient *R* values as a function of incident angle on a steel-air boundary are presented in literatures [4] and 3) A_1 and A_n values can be measured with respect to a reference amplitude by comparing peak amplitudes corresponding to the first backwall on the pulse echo signals obtained from each element. Repeating the above described approach using sound paths between any 2 elements of the aperture can lead to calculating the beam profile for more angle – sound path combinations. The combination can be increased further when the array is in contact with blocks with different thicknesses. The beam profile calculated for these discrete number of angle-sound path combinations can be used to obtain beam profile for angle-sound path combinations of all pixel points in a TFM image using interpolation. The interpolated beam profile at each pixel point can then be used to create a correction factor to compensate for directivity and beam spread effects in a TFM image.

3. Proposed Calibration Procedure

The calibration process includes velocity calibration, probe delay calibration, sensitivity calibration and finally calibration for directivity and beam spread effects. A calibration block has been designed and manufactured to perform all calibration steps. The calibration block was designed to contain 6 steps with 0, 28, 46, 64, 82 and 100 mm heights as shown in Figure 6.

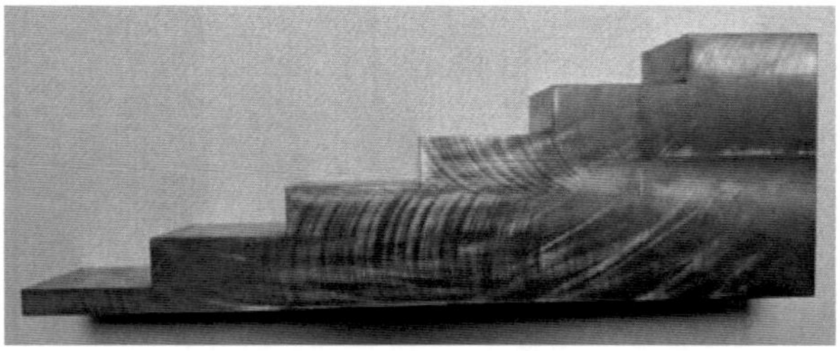

Figure 6 Step Calibration Block and Calibration Process.

The proposed calibration procedure can be summarized as follows:

1) Velocity calibration is performed by comparing travel time between two consecutive backwall echoes on A-scans captured when an element is used in transmit mode and the same element is used in receiving mode. The measured travel time is used with the block thickness to calculate velocity.
2) Probe delay calibration is required to compensate for time delay in pulsing between different elements. A-scans obtained from each element in a pulse echo mode is used for calibration. Time corresponding to the peak amplitude from first backwall reflection is recorded from each element's pulse-echo A-scan. Recorded times were compared to obtain elements pulsing time delays.
3) Sensitivity calibration is required to compensate for variation in energy transmitted by each element. The peak amplitude corresponding to the first backwall reflection is recorded from each element's pulse-echo A-scan. Peak amplitude values are normalized with respect to the maximum amplitude recorded. The normalization factor used for each element can be used to compensate for differences in energy generated by each element.
4) Calibration for directivity and beam spread is performed by calculating the beam profile generated by each array element. The probe is placed on each step of the calibration block individually and FMC data is captured. The recorded A-Scans from each step are used to calculate beam profile as explained in section 3 (using Equation 5). All angle-sound path combinations provided by the aperture length and varying block thickness that are under the aperture in each step are evaluated. Angle and sound path combinations can be converted to *x* and *y* coordinates and the measured beam profile on these coordinates is used to interpolate the beam profile on a uniform *x*-*y* grid.

A 5L64-A2 Olympus array probe (64 elements, 5 MHz center frequency, 0.6 mm pitch) was used to scan the step block shown in Figure 6. A Micropulse, 128 Element Phased Array channel PeakNDT instrument was used to collect FMC data at a sampling frequency of 25 MHz. The interpolated beam profile, *C(x,y)* obtained using this calibration procedure has been plotted in Figure 7. The beam profile has been interpolated in the regions that data points were obtained. The beam profile was found to be roughly symmetric about the element's centerline (x=0) and drops off away from the centerline and as *y* increases which is consistent with directivity and beam spread effects.

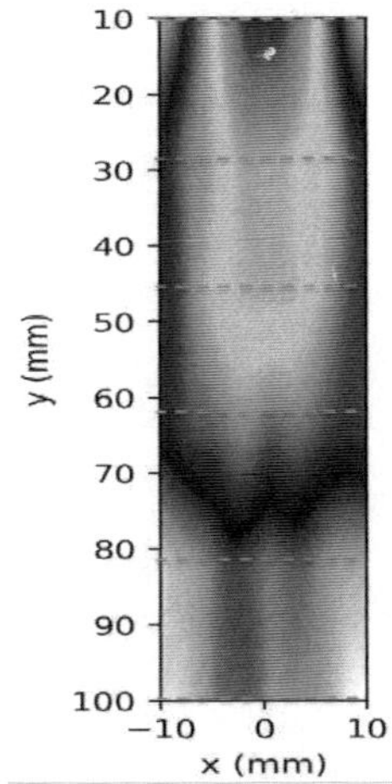

Figure 7 Beam Profile Interpolated on a uniform x-y grid.

4. Beam Profile Based Correction Factor

The beam profile obtained through the calibration process explained in the previous sections has been used to represent a correction factor which accounts for the beam profile at each pixel point of a TFM image and compensate for image intensity changes. According to Figure 8, more correction is required to be applied to image intensity at regions where beam profile drops (green and blue color mapped regions in figure 8) as a result of directivity and beam spread effects. On the other hand, less correction is needed for regions where the beam profile is stronger (red and yellow color mapped regions in figure 8). Therefore the correction factor has been introduced as the inverse of the beam profile, *1/C(x,y)* , and TFM stacking equation presented in Equation 1 can be modified to the following equation:

$$I(x,y) = \sum_{m=1}^{N}\sum_{n=1}^{N} \frac{1}{C(x,y)} A_{mn}\left(\left(d_{m(x,y)} + d_{n(x,y)}\right)/V\right) \qquad \text{Eq. 12}$$

Using Equation 12, each pixel intensity is first corrected based on the beam profile at the location of the pixel, before being used in the stacking process. For imaging points outside the range where correction data was measured, the closest correction value was employed.

5. Correction Factor Evaluation

The effectiveness of the proposed amplitude correction procedure was evaluated by imaging 3 side drilled holes (2.5 mm diameter), at various depths and horizontal positions with respect to the center of the probe aperture. An Olympus 5L64-A2 probe (64 elements, 5 MHz center frequency, 0.6 mm pitch) was used for the calibration process. For each hole, FMC captures were recorded with the center of the aperture positioned -18.9 mm, -9.45 mm, 0 mm, 9.45 mm and 18.9 mm, with respect to the horizontal position of the target. Two TFM images were then rendered from each capture; first using the regular (uncorrected) method as described by equation 1 and then using the corrected method as described by equation 12. Figure 8 present rendered TFM images for side drilled hole with a depth of 54 mm. Un-Calibrated TFM images are presented in the left column and the Calibrated TFM images are presented in the right column on the images.

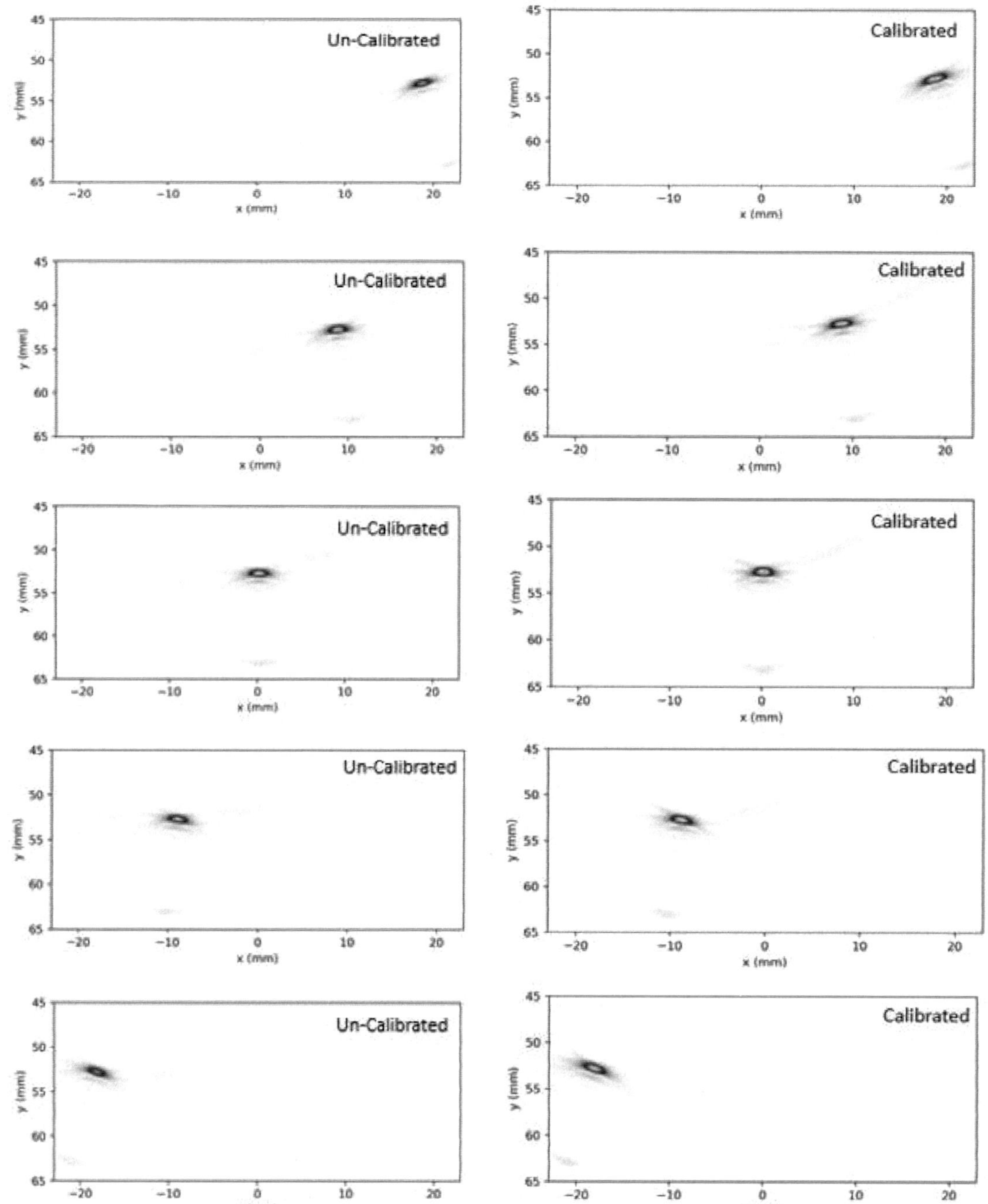

Figure 8 Un-Calibrated (Left) and Calibrated (Right) Images of the 54-mm Deep Side Drilled Hole at Different locations with respect to the Center of Aperture.

The calibration procedure equalizes the hole amplitudes, though not uniformly over the entire imaging domain. Additionally, it is noted that noise and other imaging artifacts appear brighter in the corrected images, an obvious and unavoidable consequence of applying an amplitude correction procedure. The quality of the applied correction procedure is quantified in Figure 9, where the maximum hole amplitudes for both the corrected and uncorrected images are plotted as a function of position from the center of the aperture, at each depth. Phased array amplitude calibration typically requires that holes at a constant depth should be within +/- 5% of a reference sensitivity. Here all amplitudes were normalized with respect to the maximum amplitude for the corresponding depth, so that the highest amplitude reflector is set to 100% and the analogous acceptance band becomes 90% to 100%.

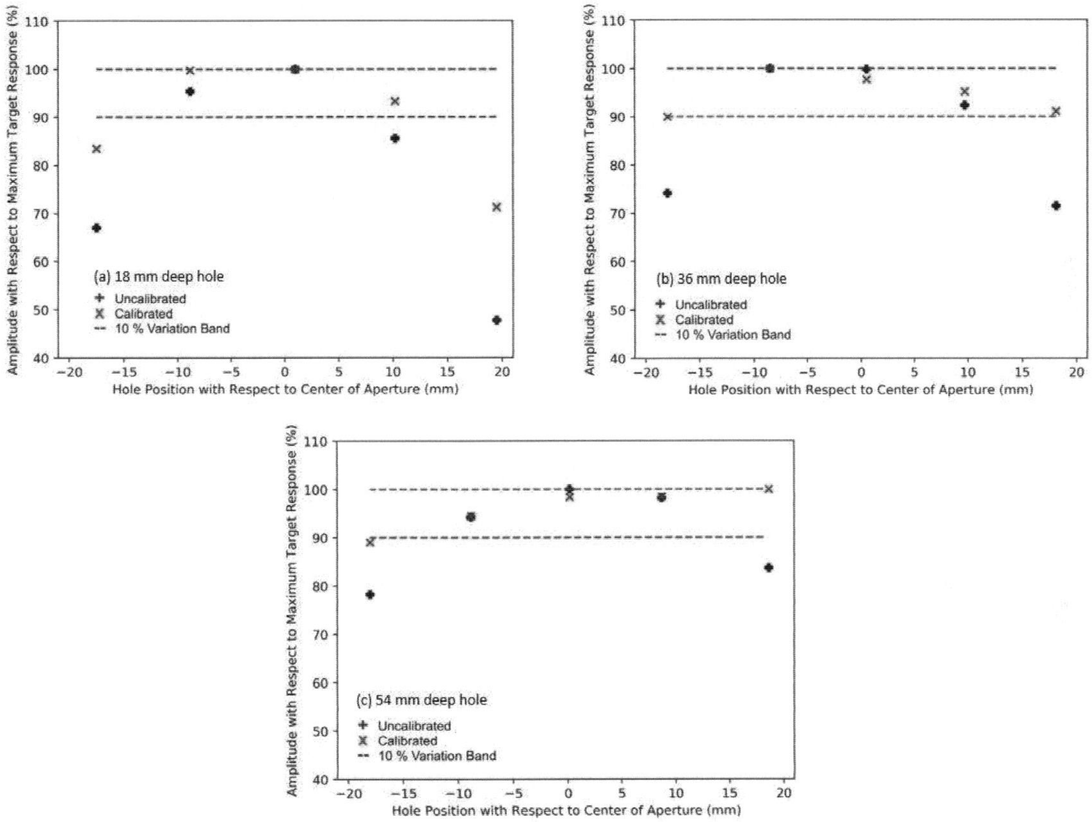

Figure 9 Max Amplitude (TFM Image Intensity) of (a) 18-mm deep, (b) 36-mm deep and (c) 54-mm deep holes as a function of position from the center of the aperture

For the shallowest hole (18 mm depth), the corrected indication amplitude was observed to be within the Phased Array code accepted [5] 10% from maximum band, for the 3 holes closest to the center of the aperture (Figure 9-a). For the 36 mm deep hole, the corrected indication amplitude was observed to be within a 10% tolerance for all positions (Figure 9-b). For the deepest hole (54 mm), the indication amplitude was observed to be within 10% of the maximum for all positions except for the hole located at -18 mm from the center of the aperture (Figure 9-c), with the measured amplitude being just below the acceptance threshold (89%). In general, the beam profile based calibration procedure seems to be more effective as the target depth increases. This is due, in part, to the fact that the difference in sound path between any 2 element pairs decreases for larger imaging depths. Nevertheless, the 10% variation in target amplitude for reflectors at a constant depth can only be met using the proposed beam profile correction scheme for targets imaged within ±10 mm of the center of the aperture, which amounts to only 53% of the total aperture length, though it is noted that only the shallowest target is out of tolerance. Clearly, the proposed procedure alone severely restricts the range for which a given probe can be acceptably calibrated based on the phased array code.

The TFM can be interpreted as synthetically point focusing at every pixel point within a chosen domain [1]. For points close to the aperture, an incredibly narrow, high intensity focus is achieved. For points far away from the aperture, the focal point becomes more spread out, with a lower maximum amplitude as shown in Figure 10. The spread of the synthetic focus is called a point spread function, and causes distortion of targets not well covered by the probe aperture; as observed in Figure 8, where the images of holes farthest from the center of the aperture appear increasingly elongated. The effects of imaging using a finite aperture cannot be neglected in correcting image amplitudes without significantly reducing the coverage range of a given phased array probe. In addition, the apparent distortion of targets near the edge of the aperture clearly implies that different sizing criteria will be required depending on where the target appears.

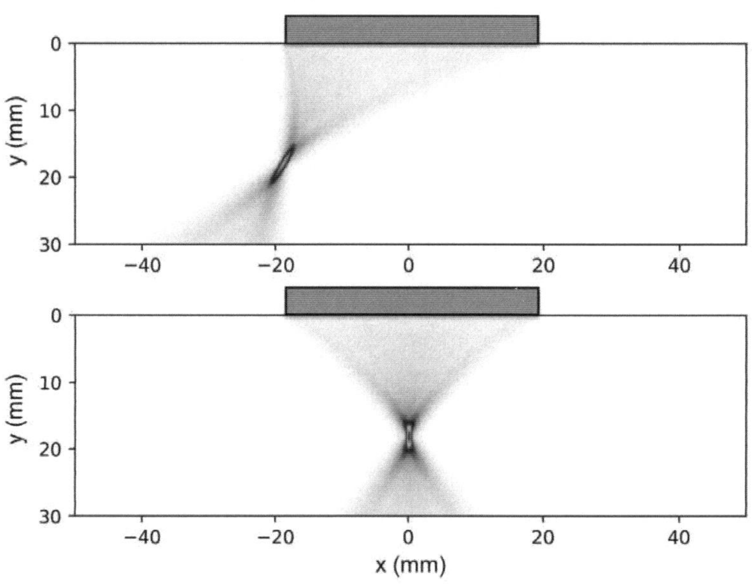

Figure 10 Effects of Focusing Away from the Center of the Aperture.

6. Conclusion

A simple procedure for estimating representative beam profiles for phased array probe elements was presented and used to equalize amplitudes across rendered TFM images. The proposed method has the advantage of estimating the amplitude corrections using measured A-scan data collected from a simple step block and does not rely on standard mathematically expressions for element field profiles which neglect the effects of probe bandwidth and passive aperture size. The technique was shown to meet the standard Phased Array code requirement of having target amplitudes within 10% of one another across the imaging range, for targets at constant depth – though this could only be achieved on the shallowest hole when imaged within 20 mm (53%) of the center of the probe aperture. Failure to calibrate amplitudes across the entire range was attributed to the effects of imaging using a finite aperture, which causes targets away from the centerline of the probe to appear distorted and at lower amplitude. The spatially varying point spread function which is intrinsic to the TFM, will have to be directly accounted for when calibrating amplitude in images rendered with the TFM, otherwise the inspection range over which amplitude can be acceptably calibrated must be roughly limited to the center 50% of the aperture, even if all elements are used.

References

[1] A J. Hunter, B.W. Drinkwater, P. D. Wilcox, "The Wavenumber Algorithm for Full-Matrix Imaging Using and Ultrasonic Array," *NDT Int.*, vol. 39, no. 7, pp. 525– 541, 2006.

[2] A. Velichko, P. D. Wilcox, B.W. Drinkwater, "Array Imaging and Defect Characterization Using Post-processing Approaches," *Advances in Acoustic Microscopy and High Resolution Imaging, from Principles to Applications,* Chapter 10. pp. 275– 318, 2013.

[3] B. W. Drinkwater and P. D. Wilcox, "Ultrasonic arrays for non- destructive evaluation: A review," *NDT Int.*, vol. 39, no. 7, pp. 525– 541, 2006.

[4] D. Cheeke, "Fundamentals and Applications of Ultrasonic Waves", CRC Series in Pure and Applied Physics, 2002.

[5] ASME 2010, Section V T-464.1.3

The effect of higher harmonic components on MPI process

Pavel Staněk, Zbyněk Škvor

Department of Electromagnetic Field
Czech Technical University in Prague

pavel.stanek@fel.cvut.cz

CINDE CANADIAN INSTITUTE FOR NON-DESTRUCTIVE EVALUATION

ELMAG.ORG

NDT in Canada NDTiC 2018 Canada's NDT Conference | June 19 - 21 Halifax Convention Centre Halifax, Nova Scotia

Content

MPI Principle

MPI steps:

- Generate magnetic field perpedicular to the defect (ideal case)
- (Verify the generated field)
- Apply detection particles
- Evaluate the indications

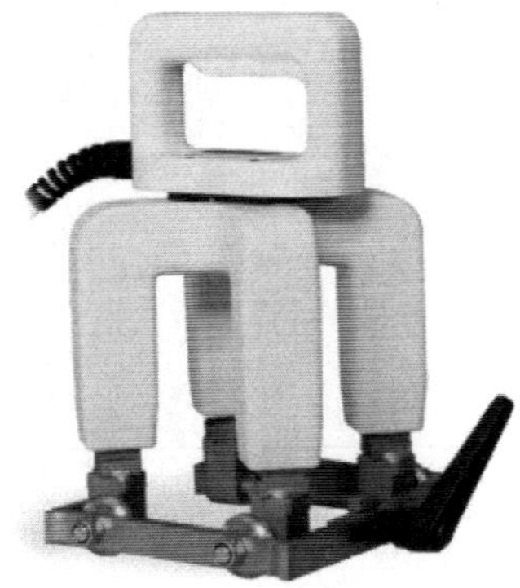

Retrieved from: http://www.sssndt.com and http://shopndt.eu

MPI Principle

MPI Principle

MPI steps:

- Generate magnetic field perpedicular to the defect (ideal case)
- (Verify the generated field)
- Apply detection particles
- Evaluate the indications

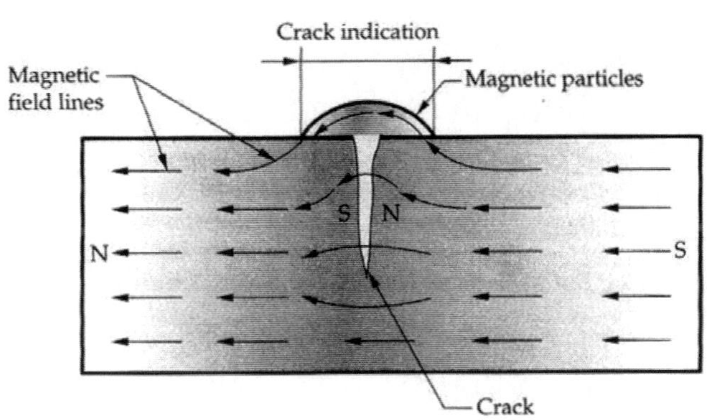

Retrieved from: http://me.aut.ac.ir

MPI Principle

MPI steps:

- Generate magnetic field perpedicular to the defect (ideal case)
- (Verify the generated field)
- Apply detection particles
- Evaluate the indications

Retrieved from: http://accutest-labs.com

Magnetic field verification

- Specimens with real well known defects
- Measurement of tangential magnetic component by a gaussmeter
- QQI - Quantitive Quality Indicators

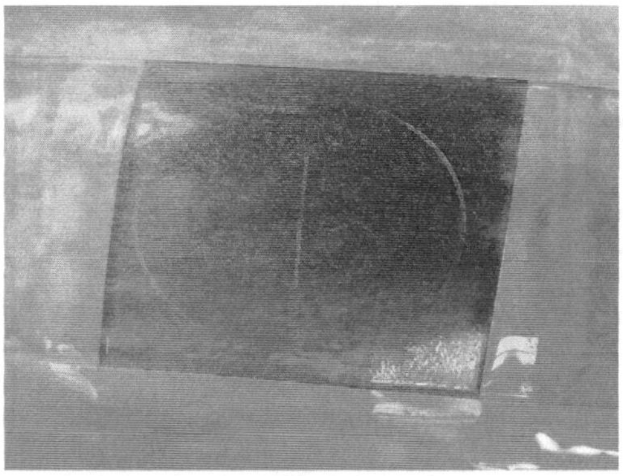

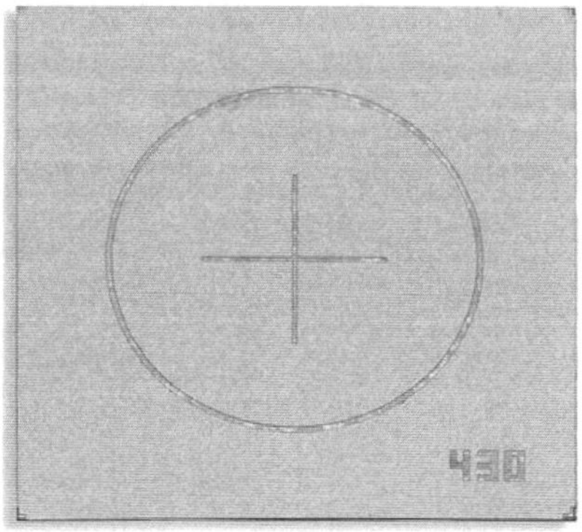

[0] Retrieved from: http://www.qnde.ca

Motivation

How to verify the field?

Need for a new method, which is

- fast
- clean
- suitable for automation

→ Measure the field and calculate the impulse of the force

Impulse of the force

Definition:

$$I = \int_{t_1}^{t_2} \mathbf{F}(\mathbf{H})dt$$

Why impulse of the force?

It combines the most critical parameters of the test

- the force exert on a detection particle
- time of duration

We need to know the leakage field above the defect and to calculate the force moving the particles.

Force impulse

The leakage field components H_x, H_y and the force components F_x, F_y were calculated by Edwards and Palmer [1]

$$F_x = -\frac{8}{3}\mu_0\pi r^3 H_0^2\left(\frac{\partial \bar{H}_x}{\partial x} + H_x\frac{\partial \bar{H}_x}{\partial x} + H_y\frac{\partial \bar{H}_y}{\partial x}\right) = k(x, y, \mu_0, \mu, a, b, r)H_0^2$$

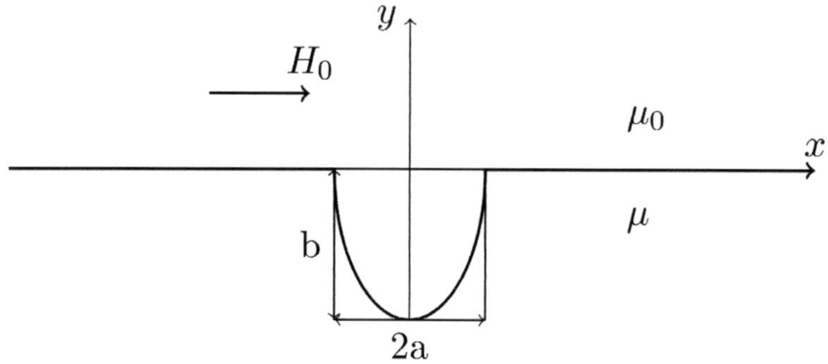

[1] Edwards, C., and S. B. Palmer. "The magnetic leakage field of surface-breaking cracks." *Journal of Physics D: Applied Physics 19.4 (1986)*: 657.

Magnetic field polarization

Example of parametric description of the polarization:

$$H_x = A cos(\omega t)$$

$$H_y = B sin(\omega t)$$

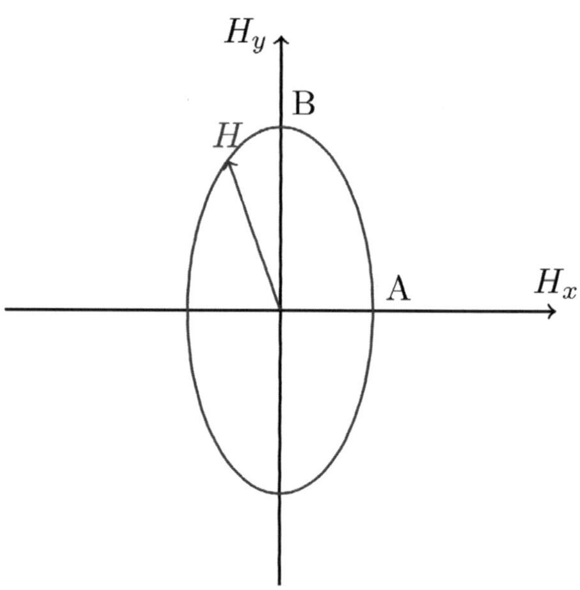

Magnetic field polarization

Field functions are continuos and periodic
→ describe them by Fourier series

$$\mathbf{H} = \mathbf{H}(H_x, H_y)$$

$$H_x = \frac{a_0}{2} + \sum_{n=1}^{\infty}(a_n cos(n\omega t) + b_n sin(n\omega t))$$

$$H_y = \frac{c_0}{2} + \sum_{n=1}^{\infty}(c_n cos(n\omega t) + d_n sin(n\omega t))$$

Magnetic field polarization

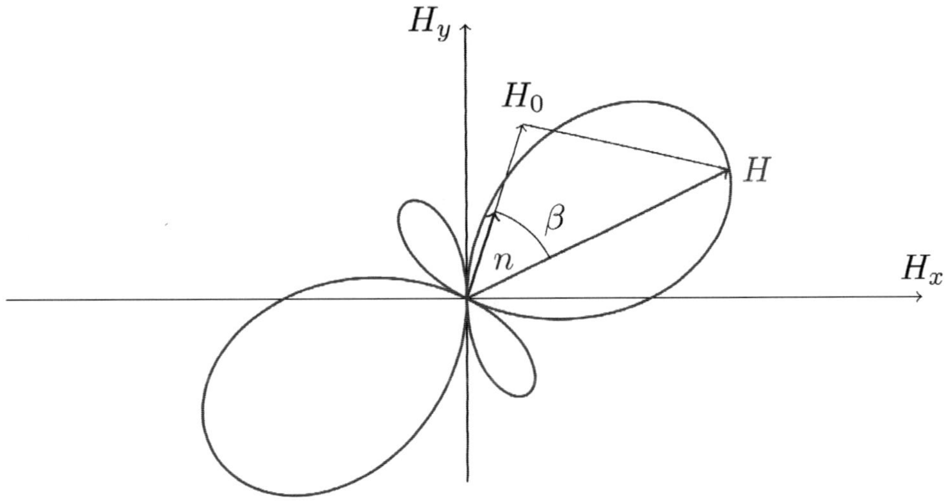

$$H_0 = |\mathbf{H}|cos(\beta) = |\mathbf{H}|\frac{\mathbf{H}\cdot\mathbf{n}}{|\mathbf{H}|\cdot|\mathbf{n}|} = \mathbf{H}\cdot\mathbf{n}$$

Magnetic field polarization measument[2]

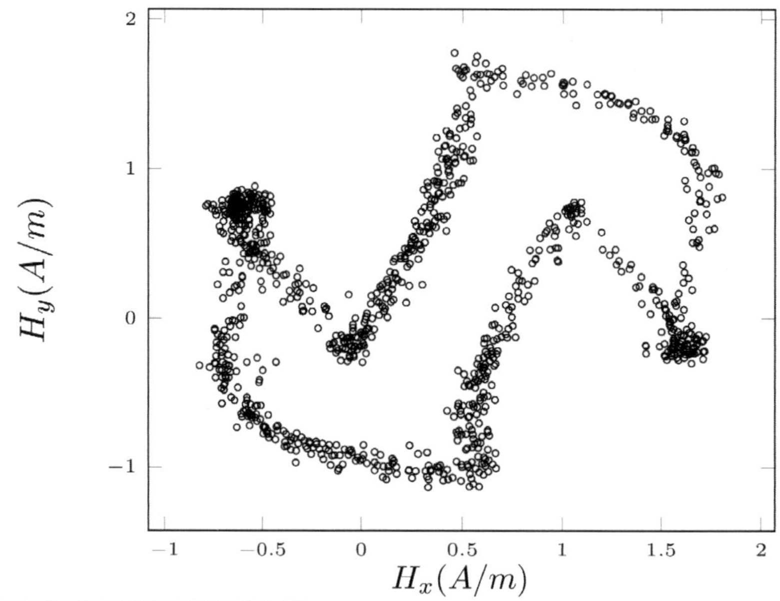

[2] Staněk, P., Škvor Z. "Experimental gaussmeter for circular magnetization." *NDT in PROGRESS, IXth International Workshop of NDT Experts, Proceedings*: Prague, 2017. ISBN 978-80-87012-63-5.

Impulse of the force

$$I = \int_0^T \mathbf{F(H)}dt = \int_0^T kH_0^2 dt =$$

$$= \frac{k}{2f}\left[cos^2(\alpha)\left(\sum_{n=1}^{N}(a_n^2 + b_n^2 - c_n^2 - d_n^2) + \frac{a_0^2}{2} - \frac{c_0^2}{2}\right) + \right.$$

$$\left. +sin(2\alpha)\left(\sum_{n=1}^{N}(a_n c_n - b_n d_n) + \frac{a_0 c_0}{4}\right) + \sum_{n=1}^{N}(c_n^2 + d_n^2) + \frac{c_0^2}{2}\right]$$

Impulse of the force

The impuslse of the force terms

$$cos^2(\alpha)\left(\sum_{n=1}^{N}(a_n^2+b_n^2-c_n^2-d_n^2)+\frac{a_0^2}{2}-\frac{c_0^2}{2}\right)$$

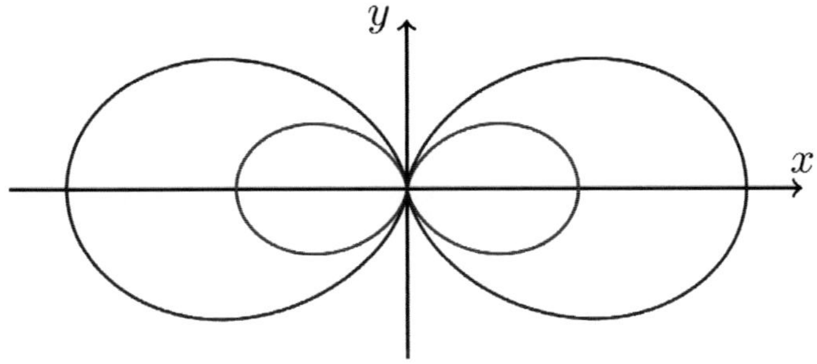

Impulse of the force

The impuslse of the force terms

$$sin(2\alpha)\left(\sum_{n=1}^{N}(a_nc_n-b_nd_n)+\frac{a_0c_0}{4}\right)$$

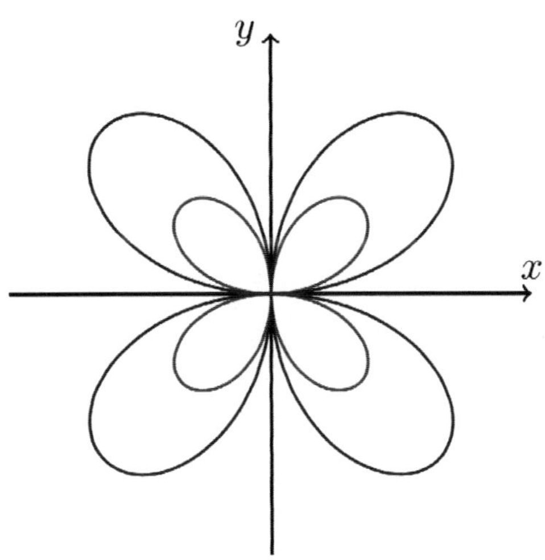

Impulse of the force

The impuslse of the force terms

$$\sum_{n=1}^{N}(c_n^2 + d_n^2 + \frac{c_0^2}{2})$$

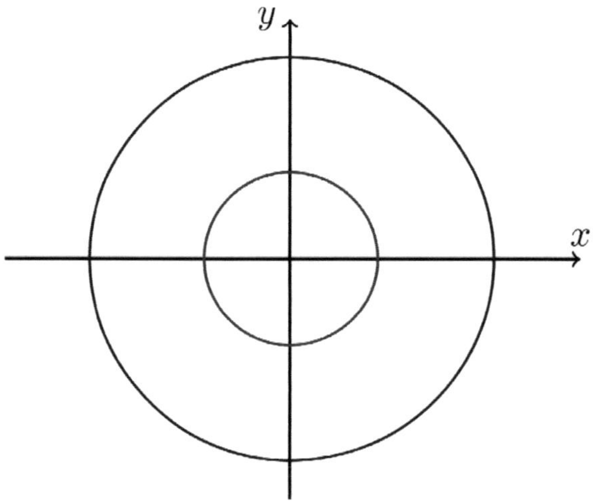

Impulse of the force

Circular polarization

Polarization

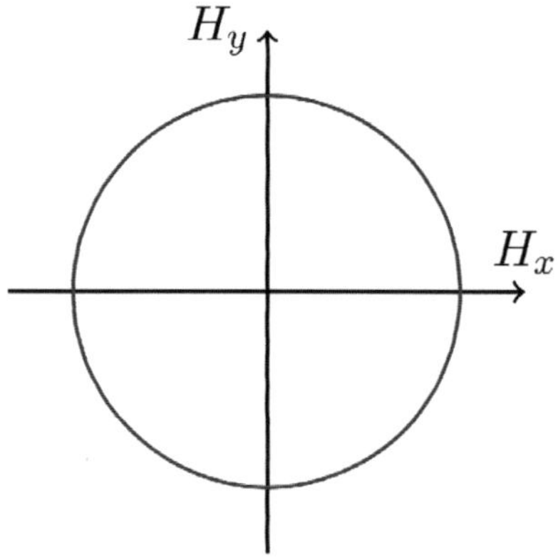

Impulse

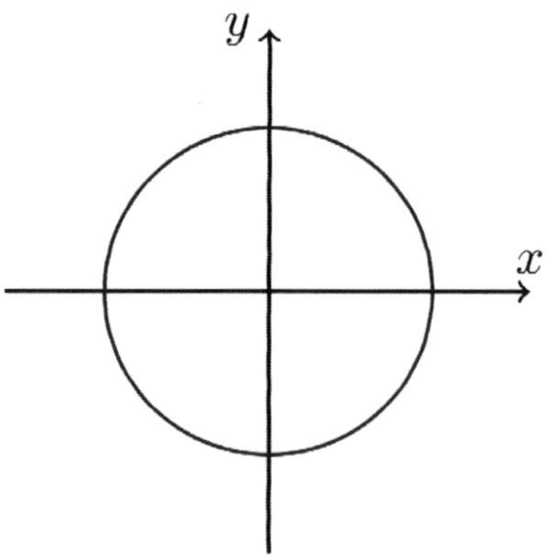

Impulse of the force

Elliptical polarization

Polarization

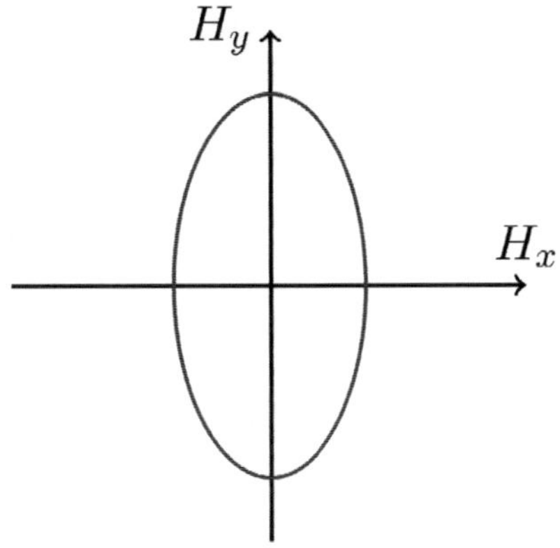

Impulse

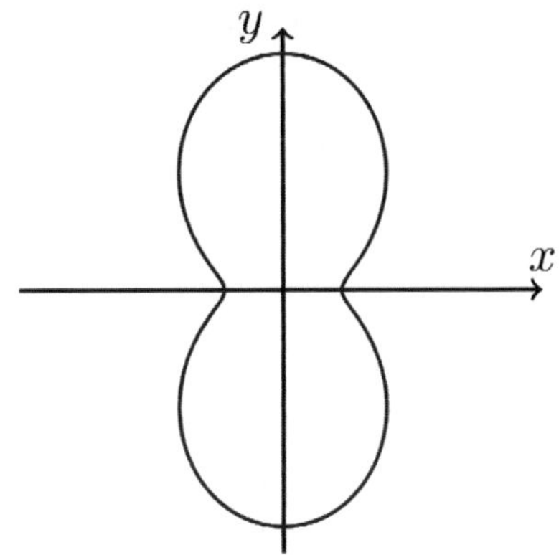

Thank you for attention

Questions?

Nickel Based Alloy IGA

Frédéric Girard [1]

[1] NDT Specialist, Pratt and Whitney Canada; Longueuil, Canada
Phone: +450-677-9411; e-mail: frederic.a.girard@pwc.ca

Abstract

Nickel based alloys are utilized in gas turbine engines as they exhibit a combination of high temperature strength, toughness and resistance to degradation in corrosive or oxidizing environments. In some turbine disc manufacturing processes, in order to prepare the component surface for Fluorescent Penetrant Inspection (FPI), a pre-penetrant etch is used. The process solution, temperature and time must all be considered when selecting a pre-penetrant etch process. If not tested for alloy compatibility the potential for IGA can go undetected and eventually lead to component failures in the field.

Keywords: Pre-penetrant etch, intergranular attack (IGA), aerospace, nickel

1. Introduction

Pre-penetrant etch is a routine operation in the manufacturing of many components. The operation consists in chemically of removing a few microns from surface of a component or an area of a component using an acid or alkali process solution for a certain amount of time at a set temperature. This is usually as preparation for penetrant inspection and the intent is to remove smeared material and open to the surface discontinuities that might not have been otherwise. A successful pre penetrant will removed the specified amount of material without causing harm to the part or be detrimental to its intended use.

2. Process controls

Many parameters should to be monitored and the obvious ones are controls of solution composition, temperature and immersion time.

Selection and composition of the process solution is important, it must be corrosive and remove material evenly. Preferential etch of microstructural feature such as grain boundaries must be avoided. Composition must be controlled at regular intervals and maintained within prescribed values.

Control of temperature must also not be neglected. For many of the etchants, higher temperature will increase reactivity. With this increase certain side effect may manifest themselves and be detrimental to hardware.

Immersion or dwell time is the other base parameter. Increase in immersion or dwell is a direct increase in material removal although it's rarely in a linear function. Prolonged immersion might also be detrimental to parts either by removing more material than necessary or by promoting selective attack.

Other factors and variables can be as important. Specimen preparation and surface finish are also factor to control. Cleanliness is actually the first factor to control when performing any etching operation. Depending on the etchant an unclean surface might not etch evenly and may cause permanent cosmetic or visual defects. Cleanliness can be achieved a variety of ways such aqueous cleaning/degreasing, vapor degreasing or even grit blasting. In the latter case, this operation must also be done with care and match with the pre penetrant etching procedure to insure that the correct amount of material is removed and that blasting operation did not cause any additional quality issues. Surface finish will have an influence as rougher surfaces will have more surface exposed to the etchant.

Dissolved metal in solution will interfere with the etching process. Depending on the alloy and the solution its tolerance will vary. Normally the threshold should be set so that the solution at with the maximum dissolve metal in it will still yield an acceptable amount of material removal.

Standard testing should include stock loss, titration, and amount of dissolved metal in solution. All of which will alloy to determine process solution "health". This should be done at regular intervals based on usage to insure conformity when processing parts. Adjustments should be made to guarantee that the process solution will stay confirming between tests. Supplemental test is often required; verification on specimens of end grain pitting and intergranular attack is often seen.

Stock loss is often used as the only test method. This can be because of available resources or economic factors but it doesn't give a full representation of the solution condition. Stock loss is a known reference just as the "TAM" panel for FPI. Material removal on the panel will rarely be on for one. Balance of chemical in process solutions cannot be established. But for certain solutions it can be the only testing method available.

On most components, the expectation is to avoid or greatly limit selective attack. Required test is usually done at regular intervals, as it is a test that captures only a moment in time. Intergranular attack is one of them; this phenomenon is corrosion occurring preferentially at grain boundaries, usually with slight or negligible attack on the adjacent grains (Fig. 1).

Figure 1

IGA will not be detected at penetrant and may, long term develop into cracks. Then penetrant sensitivity will be crucial as these might be very difficult to detect because of their size and if numerous appear as background (fluorescence). Uncontrolled pre-penetrant etch can result in component failure (fig. 2).

Figure 2

Pre-penetrant etch should be "proceduralized" and contain controls, verifications and operating parameters. Criticality of this operation can be easily underestimated when it becomes a routine operation; procedures must be followed and when results adequate cannot be achieved the root cause must be investigated, even if reprocessing (without proper authorization) the part or component would yield an acceptable result. Consequences in lack of process control are numerous and potentially important.

Mechanical & Industrial Engineering
UNIVERSITY OF TORONTO

CADIPT

Dynamic Frequency-Domain Photothermal Imaging Methodologies for Non-Destructive Evaluation of Industrial Materials: A Review of the State-of-the-Art

Pantea Tavakolian
Andreas Mandelis

CINDE
CANADIAN INSTITUTE FOR NON-DESTRUCTIVE EVALUATION

NDT in Canada
NDTC 2018
Canada's NDT Conference
June 19 - 21
Halifax Convention Centre
Halifax, Nova Scotia

Outline

1. Introduction about thermography systems
2. Lock-in thermography
3. Match Filtering thermography
4. Pulsed photothermal radar
5. Enhanced truncated correlation photothermal coherence tomography (ETC-PCT)
6. Conclusions
7. Future Work

2

Mechanical & Industrial Engineering
UNIVERSITY OF TORONTO

CADIPT

Active thermography system

Step 1: Heating; In some cases a heating source controller is connected to a signal generator to receive the modulating signal.

Step 2: thermal diffusion through the material bulk following the heat deposition cut-off at the surface.

Step 3: Infrared camera detection.

Step 4: The thermogram sequence is processed with the reconstruction algorithm on a computer.

A generic active thermography instrumentation

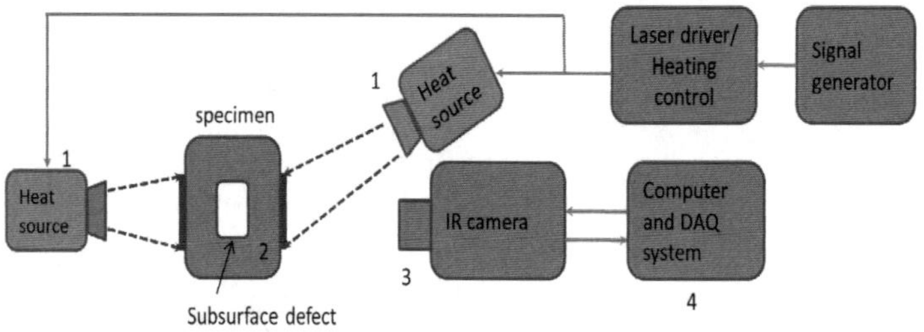

Mechanical & Industrial Engineering
UNIVERSITY OF TORONTO

CADIPT

Lock-In thermography

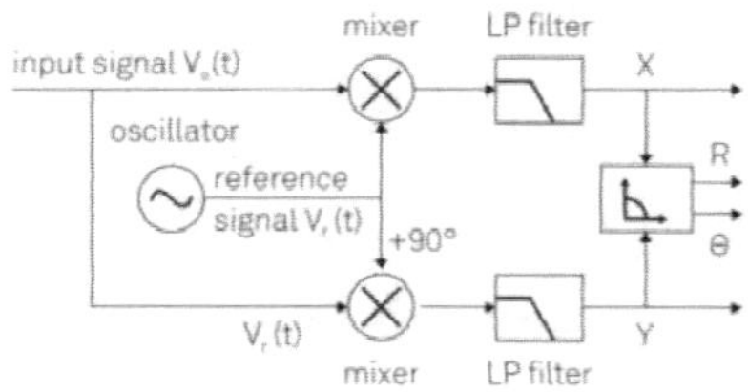

$$= \sqrt{(S^0)^2 + (S^{90})^2}$$

$$= \arctan\,(S^{90}/S^0)$$

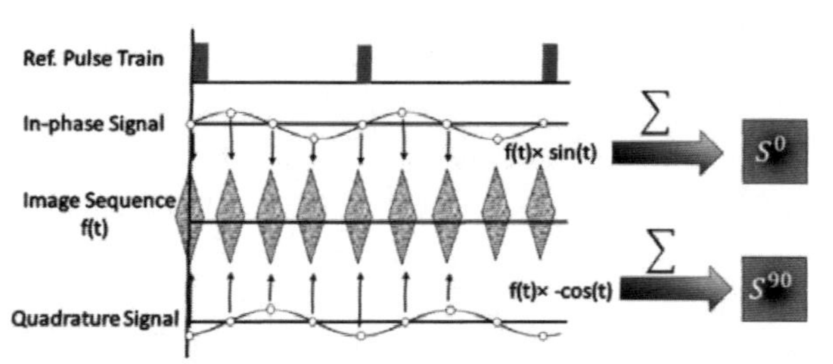

Ashkan Ojaghi, Artur Parkhimchyk, and Nima Tabatabaei, "First step toward translation of thermophotonic lock-in imaging to dentistry as an early caries detection technology" (2016)

Mechanical & Industrial Engineering
UNIVERSITY OF TORONTO

CADIPT

Match-filter thermography algorithm

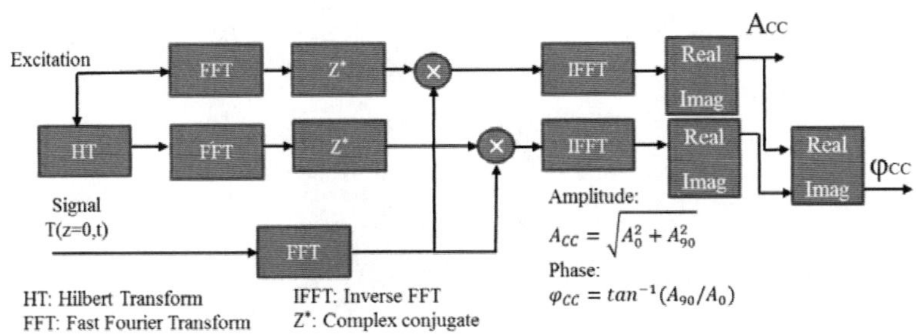

Mechanical & Industrial Engineering
UNIVERSITY OF TORONTO

CADIPT

Thermal wave-radar

A steel sample containing several blind holes with steel overlayer of various thicknesses.

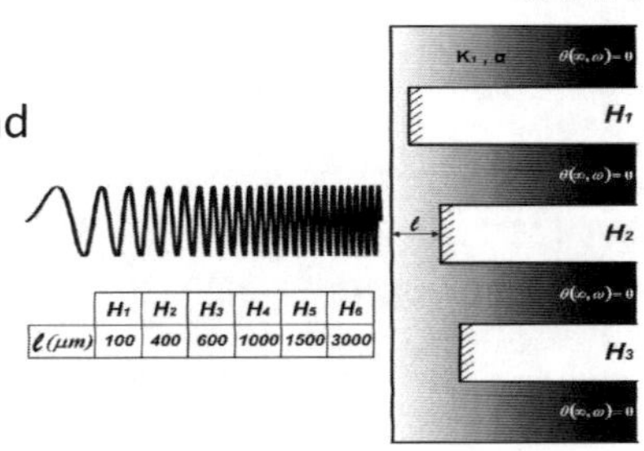

	H_1	H_2	H_3	H_4	H_5	H_6
ℓ (µm)	100	400	600	1000	1500	3000

Theoretical TWR signals for 100-µm thick and semi-infinite steel samples. The inset magnifies the time interval checked in the main plot

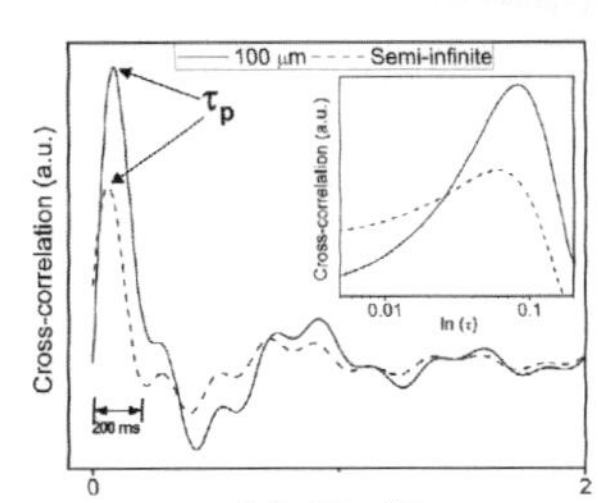

Binary Phase Coding Thermal Coherence thermography

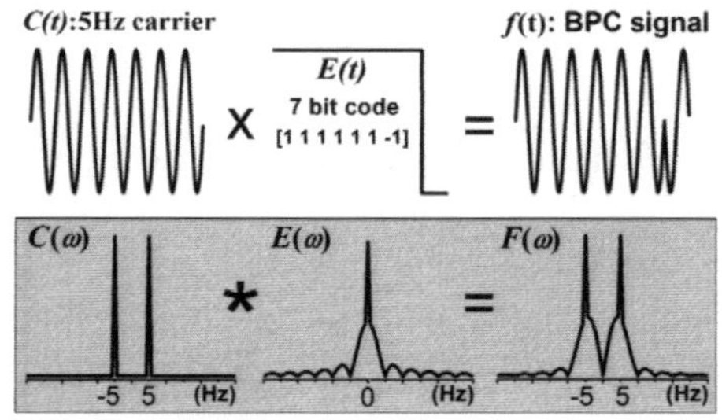

- Binary Phase Coding (BPC) signal construction in time (top) and frequency (bottom) domains for a 7-bit code and 5 Hz carrier

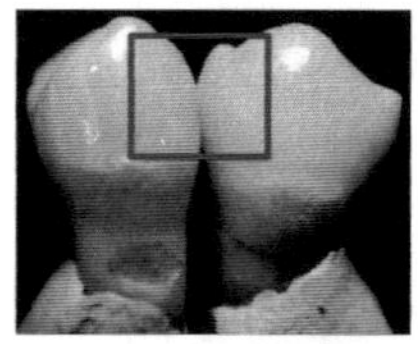

- Teeth matrix with hidden interproximal early caries. The red rectangle shows the imaged area. (b) Conventional LIT and (c) TCT phase images of the teeth matrix

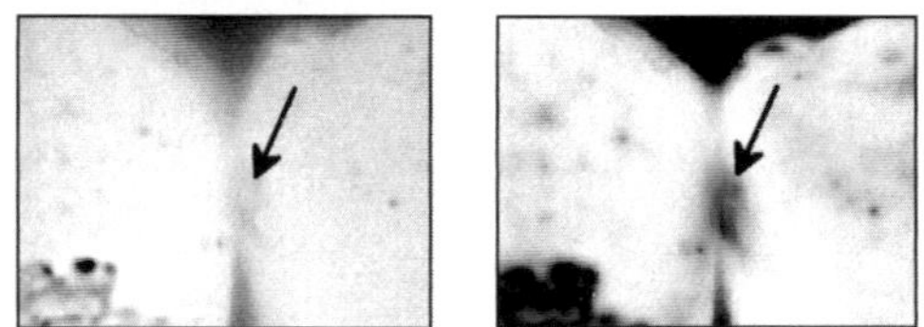

(b) (c)

Mechanical & Industrial Engineering
UNIVERSITY OF TORONTO

CADIPT

Pulsed chirp photothermal radar algorithm

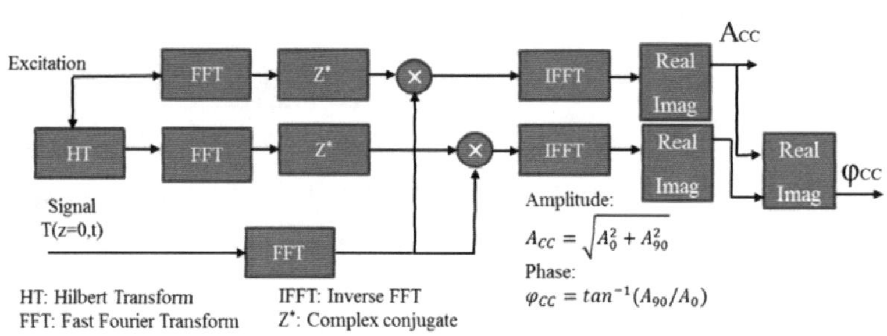

Mechanical & Industrial Engineering
UNIVERSITY OF TORONTO

CADIPT

Pulsed chirp photothermal radar

$$T(\vec{r}, t; \omega) = T_0 + T_{dc}(\vec{r}, t; \omega) + T_{ac}(\vec{r}, t; \omega)$$

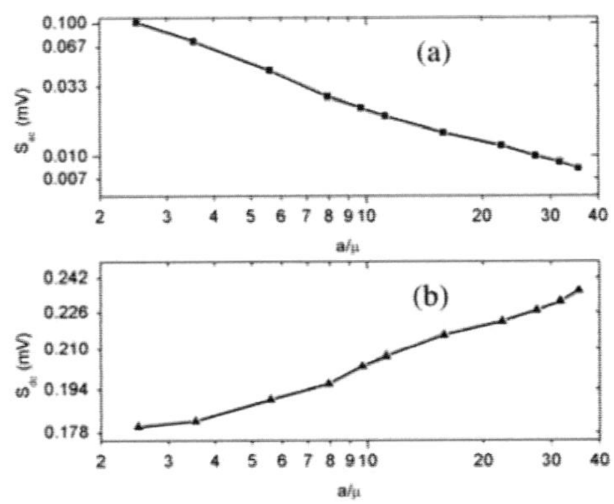

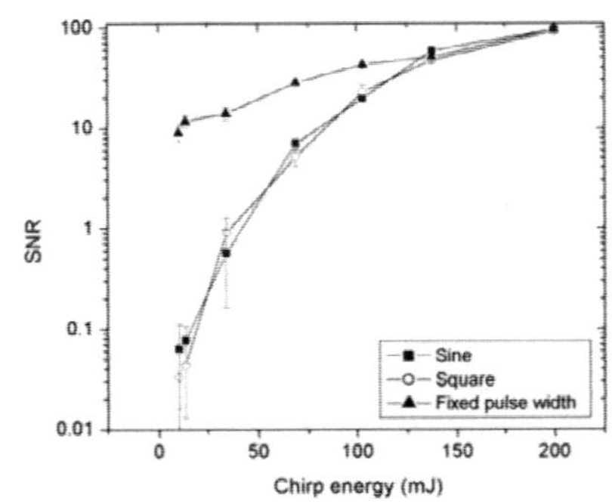

where a is the laser beam spotsize and u is the thermal diffusion length

Enhanced Truncated Correlation-Photothermal Coherence Tomography (E-TC-PCT) reconstruction algorithm

CADIPT

- E-TC-PCT cross-correlates the in-phase and the quadrature of a delay-incremented reference chirp ($R_{n,0}$ and $R_{n,90}$) with the thermal images (S) captured by the camera in a pixel-by-pixel format.

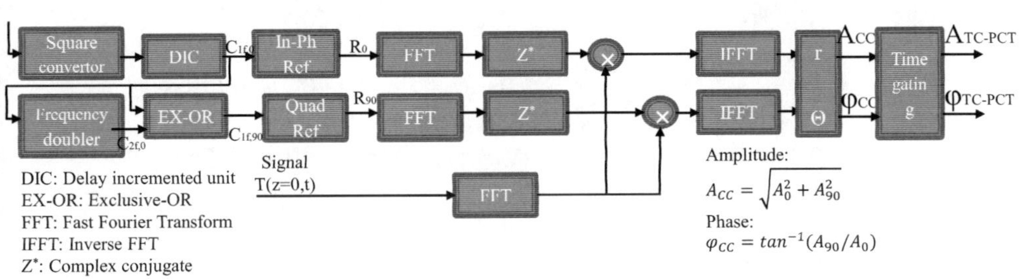

Enhanced TC-PCT reconstruction algorithm, and amplitude and phase measurements

10

Mechanical & Industrial Engineering
UNIVERSITY OF TORONTO

CADIPT

Excitation and thermal signal generation

- Excitation pulses: short duration pulses under thermal confinement (much less than the thermal relaxation time) with high peak power.
- Reference chirp: Phase incrementing (d in ms)
- A chirp is a signal in which the frequency increases (up-chirp) or decreases (down-chirp) with time.

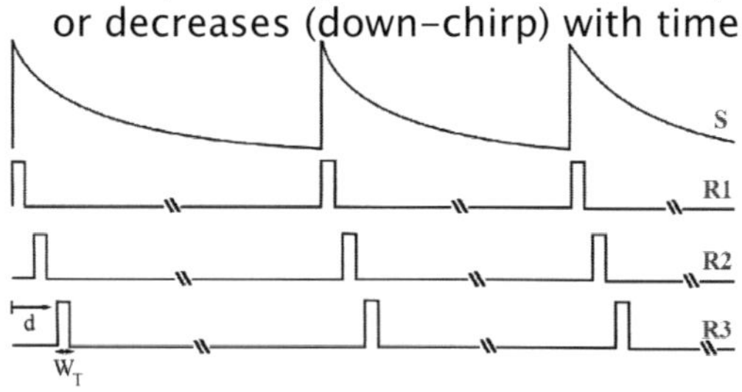

Photothermal chirp

1st reference chirp

2nd reference chirp

3rd reference chirp

NDT in Canada 2018 | June 19-21 | Halifax, NS

Steel sample with several subsurface hole

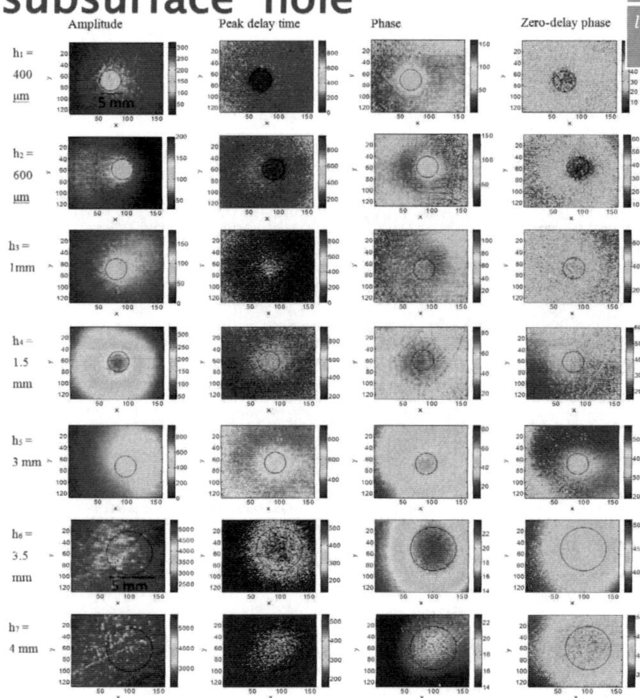

	h_1	h_2	h_3	h_4	h_5	h_6	h_7
$l(mm)$	0.4	0.6	1	1.5	3	3.5	4

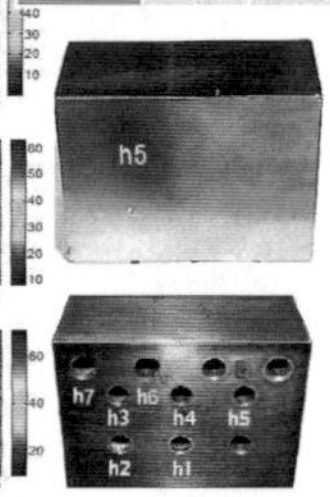

E-TC-PCT images of the holes in the steel sample. Each row displays four E-TC-PCT output channels (amplitude, peak delay time, phase, and zero-delay phase) for one of the holes.

12

3D images of holes ≤ 1.5 mm deep

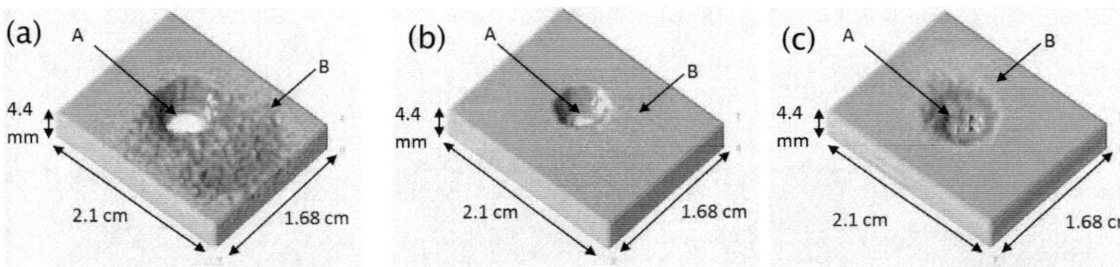

	h_1	h_2	h_3	h_4	h_5	h_6	h_7
$l(mm)$	0.4	0.6	1	1.5	3	3.5	4

3-D visualization of the invisible subsurface holes in a steel block sample using TC-PCT imaging. Amplitude tomograms of the enhanced TC-PCT of hole h_1 (a), h_2 (b), and h_3 (c) with truncation time gate of 40 ms. Each image covers an area of 2.1 cm × 1.68 cm on the sample, and the image delay time range is 1.2 s which corresponds to depth of 4.39 mm. A and B on these images display the holes, and the energy accumulation at the back surface, respectively. The laser beam illuminating the sample from the bottom is a pulsed chirped signal with sweep range of 0.2 Hz-0.6 Hz, duration 12 s, and pulse duration 10 ms. Diameter of the illumination beam is 3 cm, and the laser fluence per pulse is 7 mJ/cm².

13

Mechanical & Industrial Engineering
UNIVERSITY OF TORONTO

Lock-in thermography

CADIPT

Amplitude | Phase

h_1 = 0.4 mm & h_2 = 0.6 mm

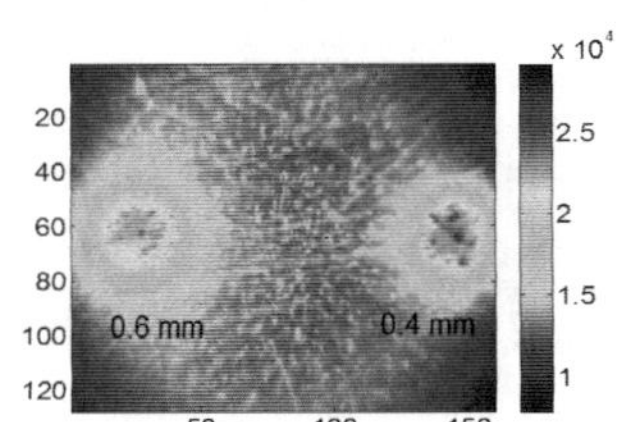

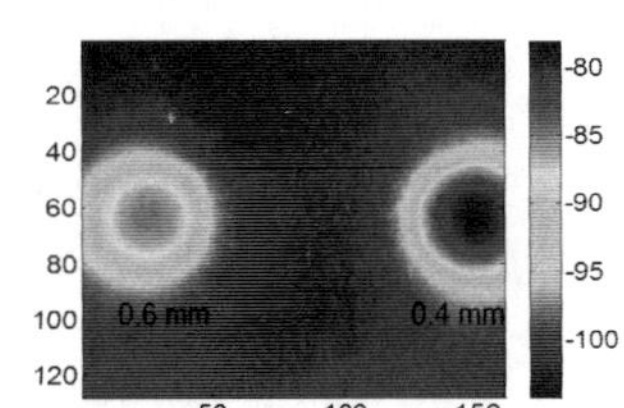

h_1 = 1 mm & h_2 = 1.5 mm

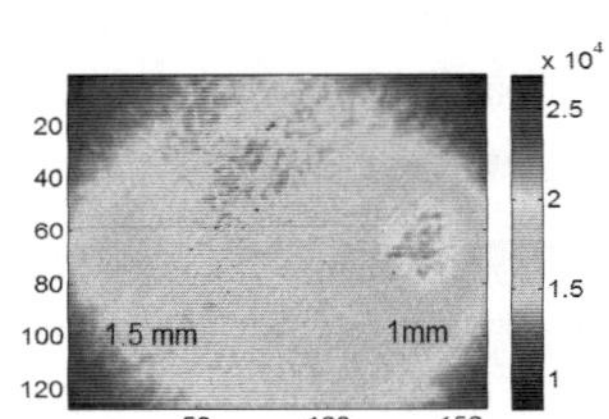

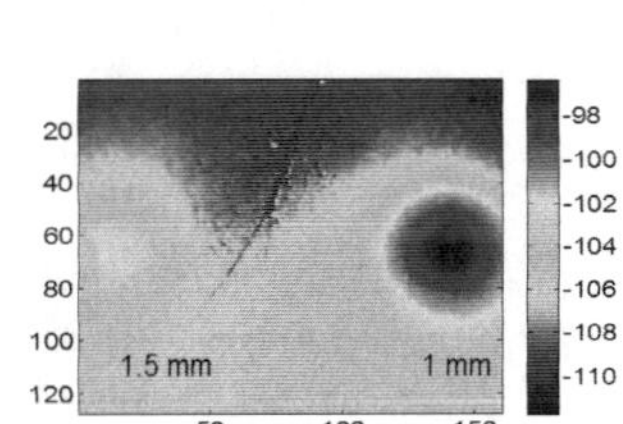

NDT in Canada 2018 | June 19-21 | Halifax, NS

14

Mechanical & Industrial Engineering
UNIVERSITY OF TORONTO

CADIPT

Thermal wave radar

Amplitude | Amplitude Delay | Phase

$h_1 = 0.4$ mm

&

$h_2 = 0.6$ mm

$h_1 = 1$ mm

&

$h_2 = 1.5$ mm

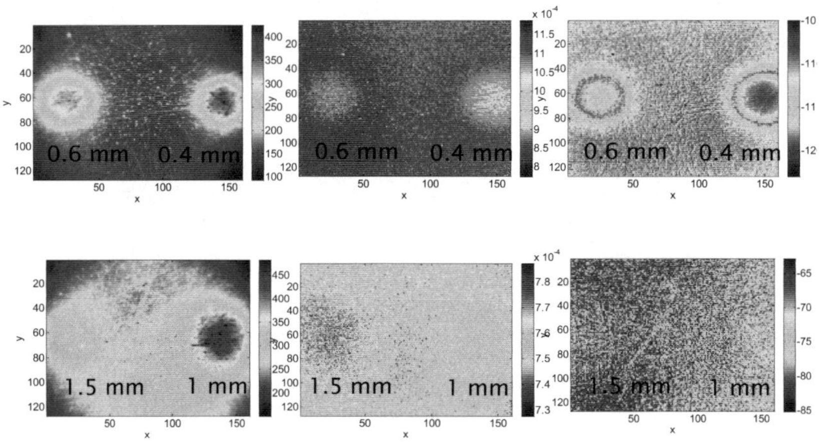

Defects in the inlay sample

A photograph of the front side (recto) of the final sample is shown in (a) and a middle layer underneath (a) is shown in (b).

The positions, names and sizes of the defects in the inlay sample: c) recto, and d) verso.

It should be noted that defect 1 affects the entire middle layer, while defect 2 affects only one half of it.

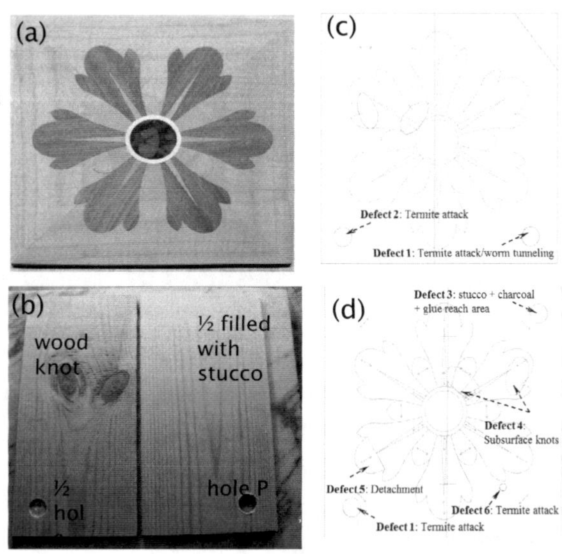

Cross-sectional image of the sample (e)

Defectoscopy in an art object

Imaging area: 3.22 cm × 2.58 cm

TC-PCT images of a delamination of a top layer

TC-PCT images of drilled hole in the middle

TC-PCT images of ½ drilled hole in the middle

TC-PCT images of a wood knot in the middle layer

17

E-TC-PCT images of a delamination of a top layer

E-TC-PCT images of drilled hole in the middle

E-TC-PCT images of ½ drilled hole in the middle

E-TC-PCT images of a wood knot in the middle layer

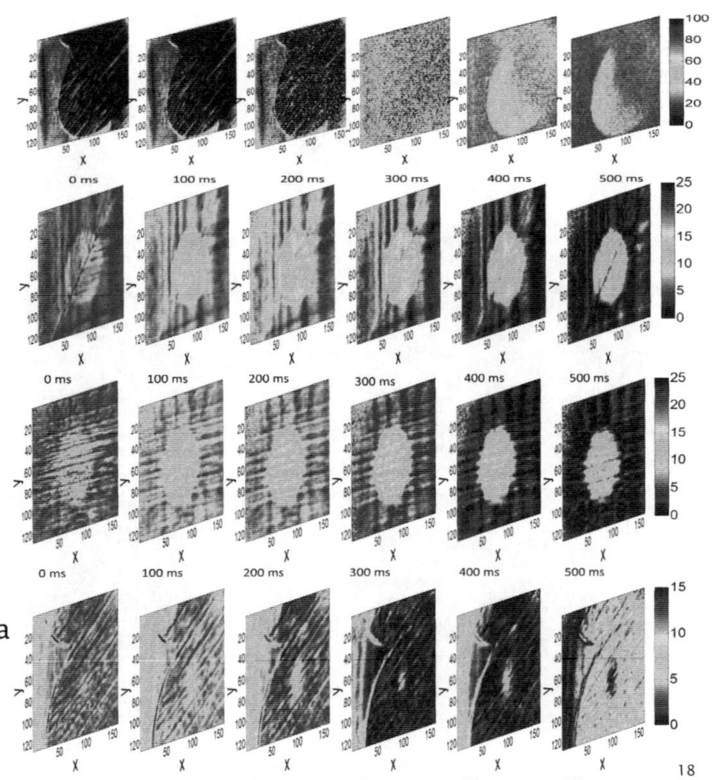

18

Defectoscopy in an art object

TC-PCT images of filled hole with stucco and charcoal in the middle

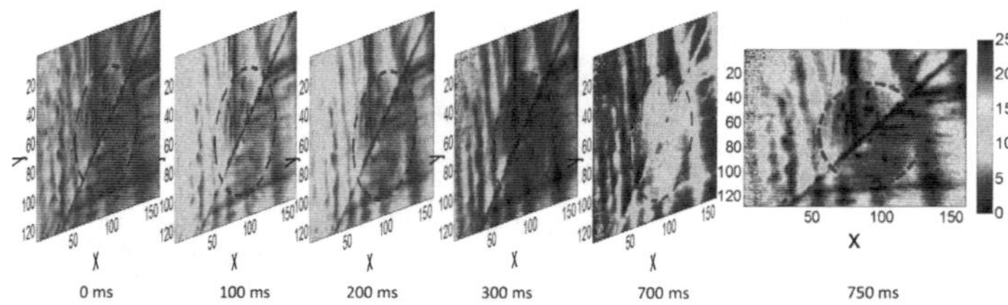

TC-PCT images of a larger area showing two defects in one image experiment

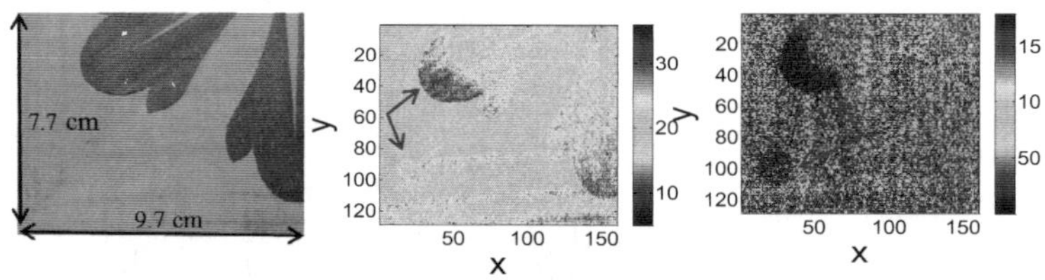

TC-PCT images of a deep hole in the middle layer

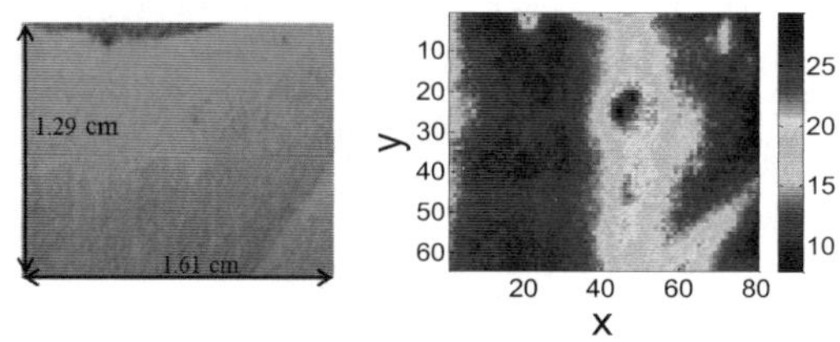

19

Mechanical & Industrial Engineering
UNIVERSITY OF TORONTO

Dental imaging

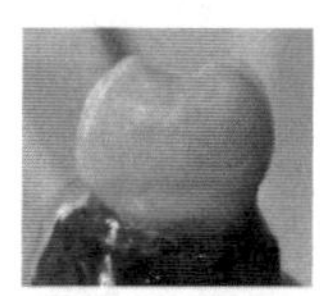

Phase

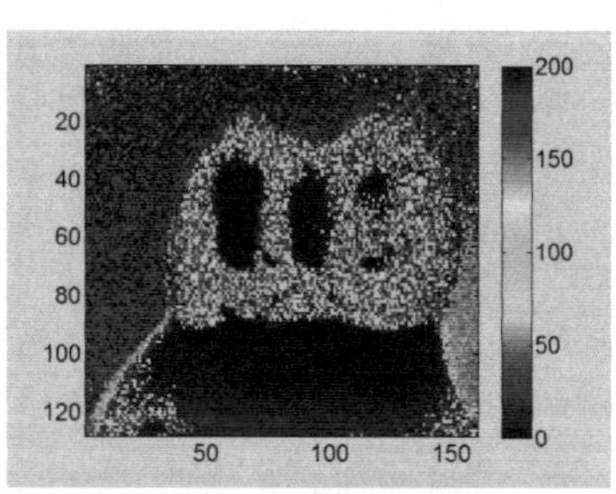

Phase delay

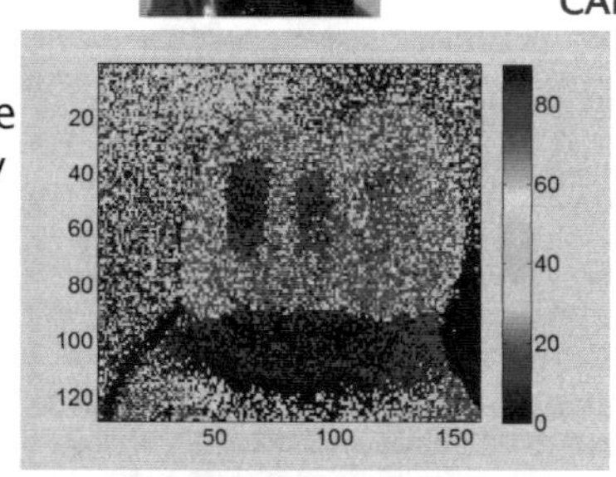

Amp

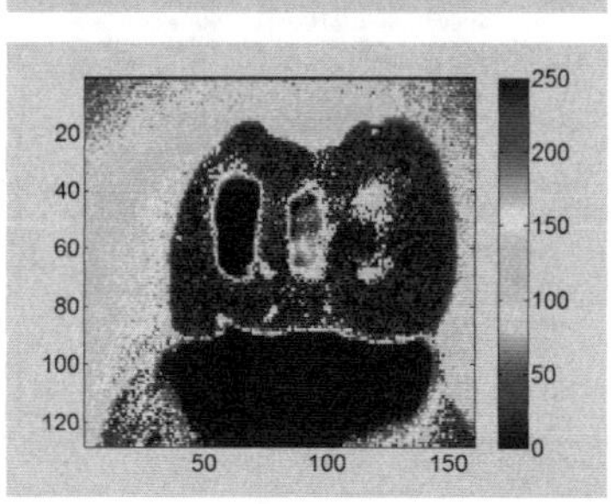

Amp delay

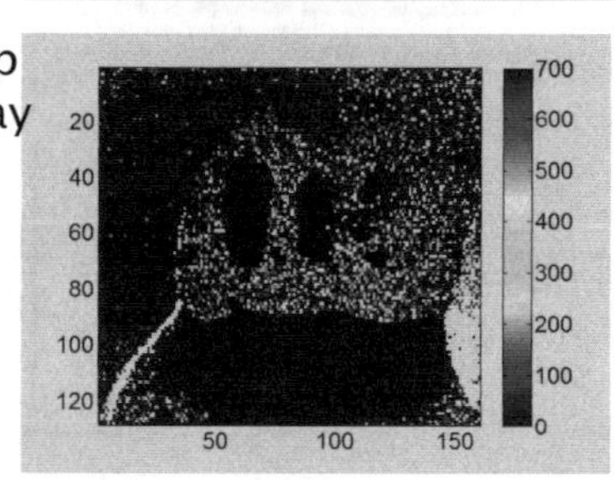

NDT in Canada 2018 | June 19-21 | Halifax, NS

Conclusions

- Enhanced TC-PCT
  - is a new thermal imaging modality that can provide high resolution 3D images of different specimen
  - has higher axial resolution and depth range compared to all other modalities
  - can be applied for non-destructive testing of industrial materials and art objects
  - is able to be used for biomedical applications

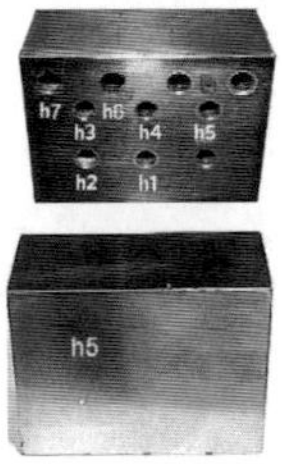

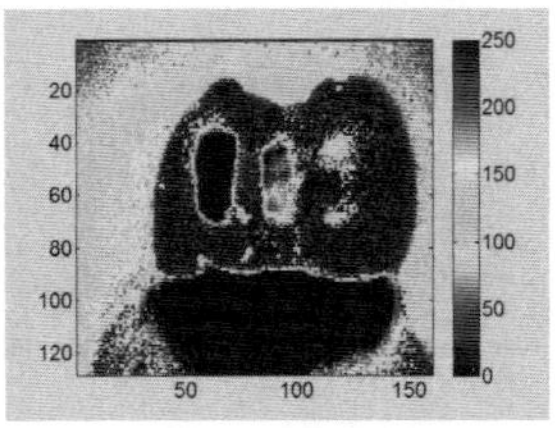

21

Future work

- Signal analysis for each application to realize an image enhancement method for different case studies.
- New applications such as diagnosis of Osteoporosis

Thank you!

SONOTEC
ULTRASONIC SOLUTIONS

A New Approach to Air-Coupled Broadband Measurement:

Effective Testing of Composite Laminates by Using A New Multi-Element Transducer

Dr. Ralf Steinhausen[1]
Dr. Mario Kiel[1]
Andreas Bodi[2]
Tobias Gautzsch[2]

[1] Forschungszentrum Ultraschall Halle gGmbH, Germany, ralf.steinhausen@fz-u.de
[2] SONOTEC Ultraschallsensorik Halle GmbH, Germany, t.gautzsch@sonotec.de

CINDE
CANADIAN INSTITUTE FOR NON-DESTRUCTIVE EVALUATION

NDT in Canada
NDTC 2018
Canada's NDT Conference
June 19 - 21
Halifax Convention Centre
Halifax, Nova Scotia

SONOTEC
ULTRASONIC SOLUTIONS

OUTLINE

- Introduction
- Motivation
- Phased-Array ACUT Probe
- Broadband Multi-Element ACUT Probe
- Application
- Test Results
- Conclusion

Introduction

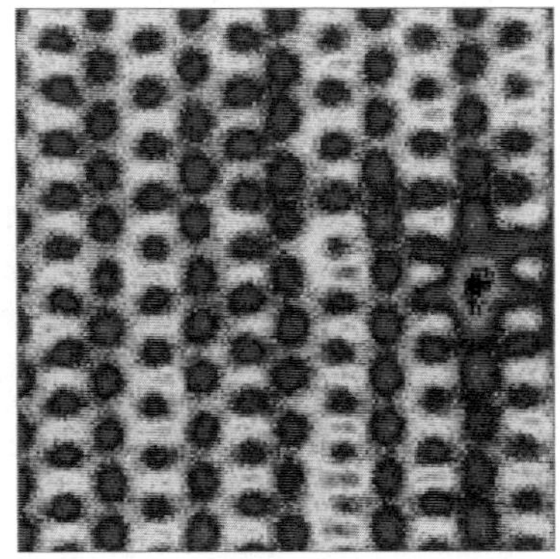

Motivation – Frequency Dependent Flaw Detection

- Test object: Composite with delamination

365 – 390 kHz | 390 – 403 kHz | 403 - 415 kHz | 415 - 428 kHz | 428 - 440 kHz

- 5 Scans
- Equal Setup
- 5 different center frequencies

→The detectability of delaminations and the outline of the flaw is frequency dependent
→High Bandwidth Transducers could lead to a better detection and sizing of delaminations

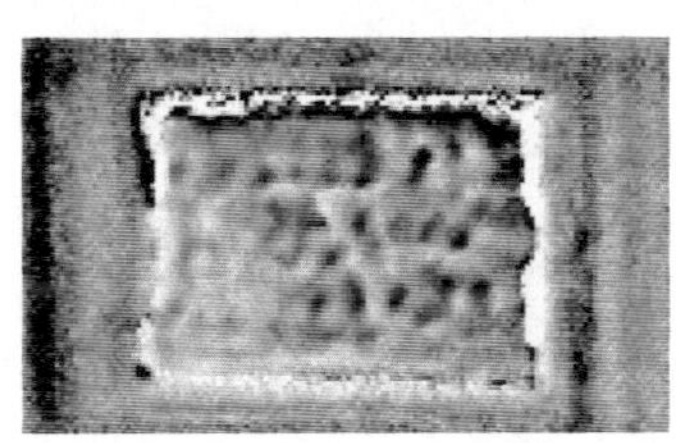

Spectral Analysis

SONOTEC
ULTRASONIC SOLUTIONS

Phased-Array ACUT Probe

- Piezocomposite Transducer
- Dice and Fill

- 3 Elements
- Equally sized
- Structured Electrode
- Annular Array

- 400 kHz
- Small Bandwidth
- Single Peak Spectrum

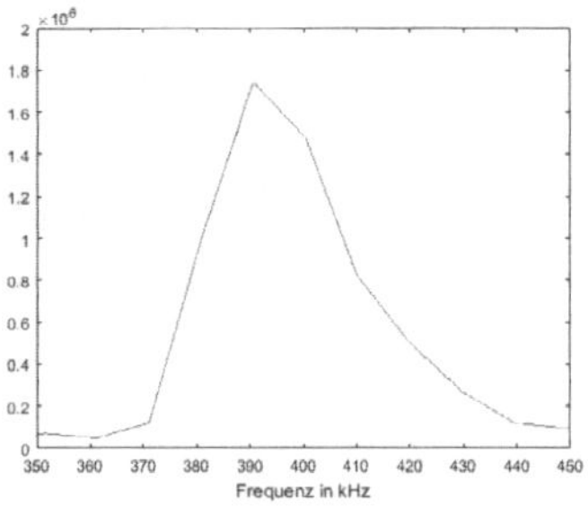

SONOTEC
ULTRASONIC SOLUTIONS

Broadband Multi-Element ACUT Probe

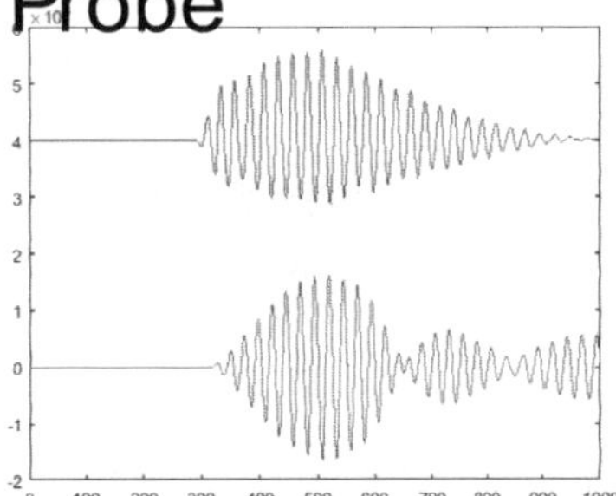

- 3 Elements
- Equally sized area
- Structured Electrode
- Annular Array

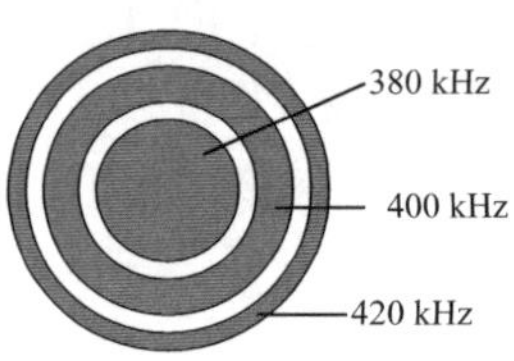

- 380, 400 and 420 kHz Elements
- Wider Bandwidth
- Three Peak Spectrum
- Shorter A-Scan Burst

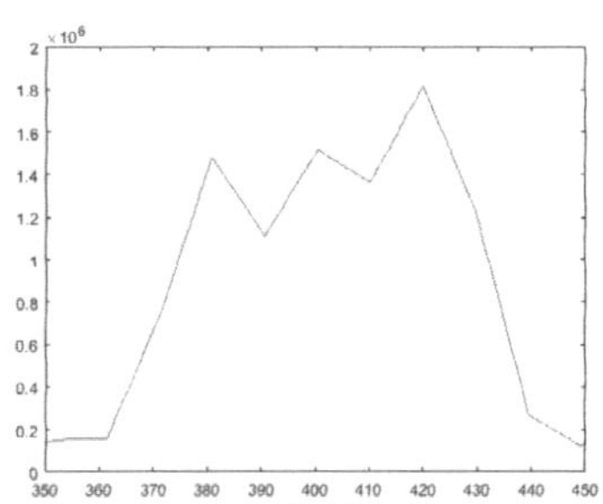

Broadband Multi-Element ACUT Probe – Test Results

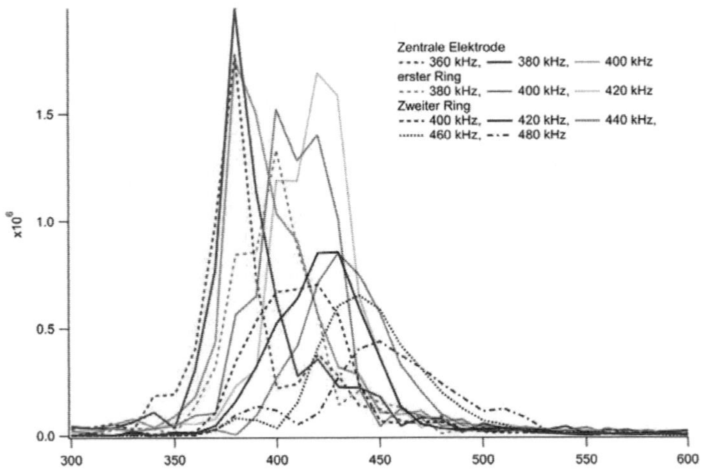

Equipment

- High End 4-Channel ACUT Electronics
  - Powerful transmitter up to 800 V (4 kW)
  - Freely programmable signal generator (50 - 750 kHz)
  - High dynamic low-noise amplifier up to 120 dB at 1 nV / $\sqrt{\text{Hz}}$
  - Customizable Software SONOSTUDIO
  - Full Data Access

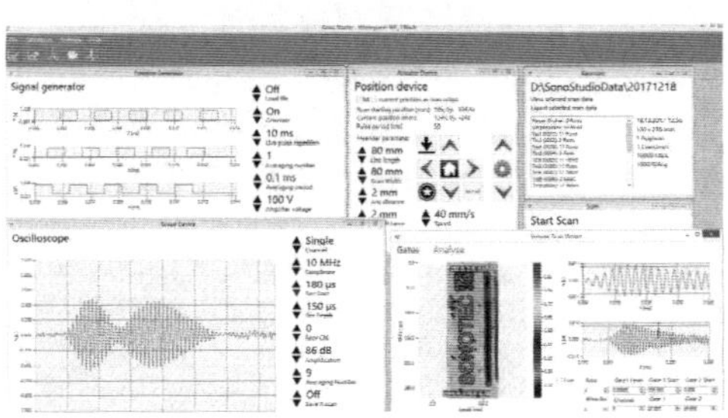

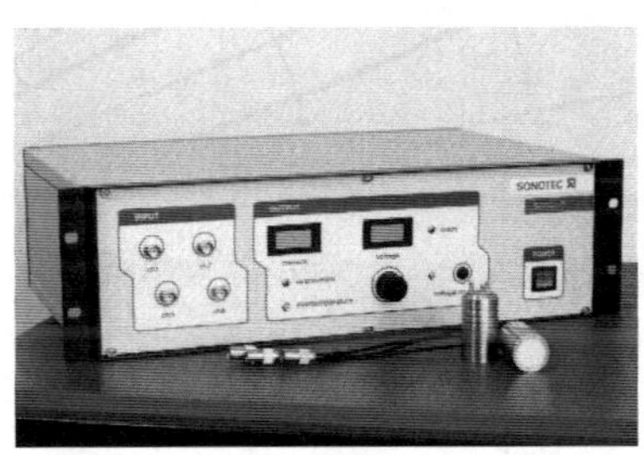

SONOTEC
ULTRASONIC SOLUTIONS

Application

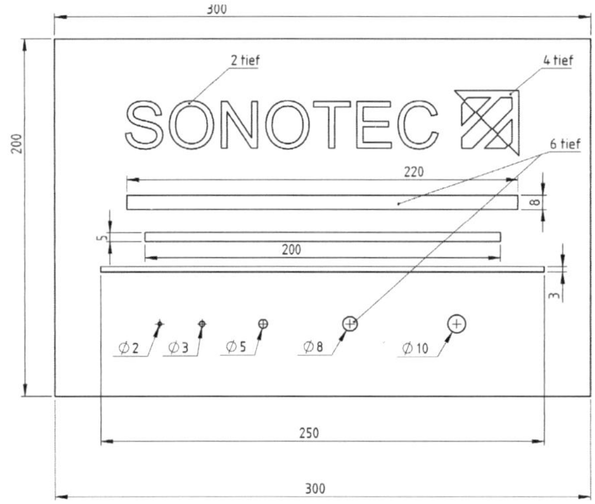

- Reference scan

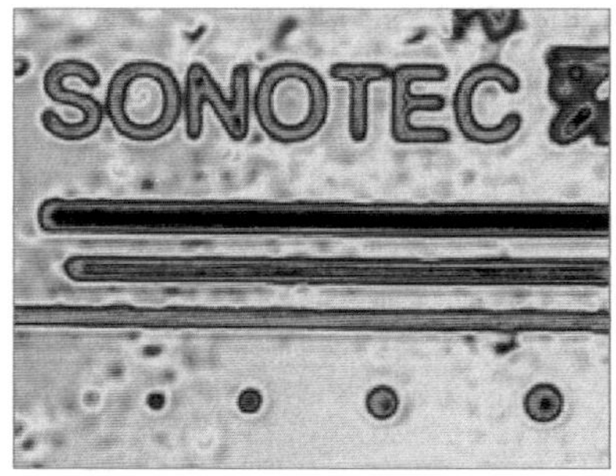

- 400 kHz through transmission
- Phased-Array transducer
- Ø2 mm detectable

SONOTEC
ULTRASONIC SOLUTIONS

Test Results - Scans

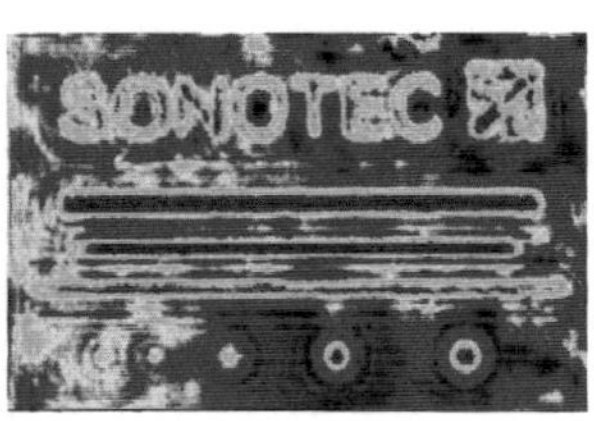

380 kHz

400 kHz

420 kHz

- Differences in Amplitude drop
- Differences in interference pattern
- Less focused than the reference scan

NDT in Canada 2018 | June 19-21 | Halifax, NS

SONOTEC
ULTRASONIC SOLUTIONS

Test Results – Spectral Analysis

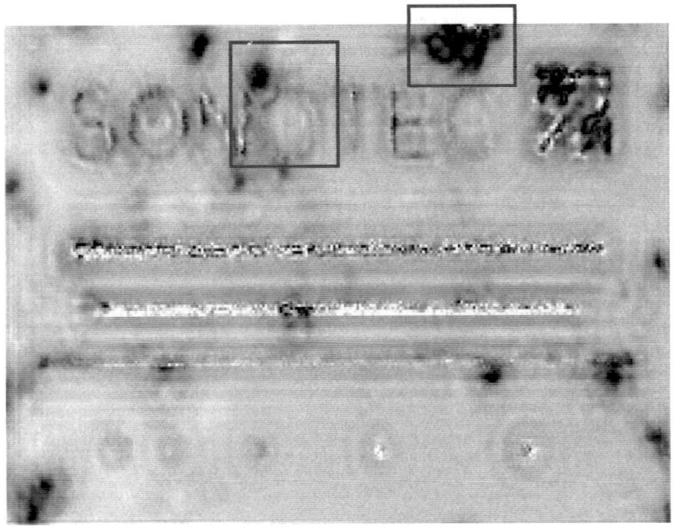

NDT in Canada 2018 | June 19-21 | Halifax, NS

SONOTEC
ULTRASONIC SOLUTIONS

Test Results – Exemplary Analysis

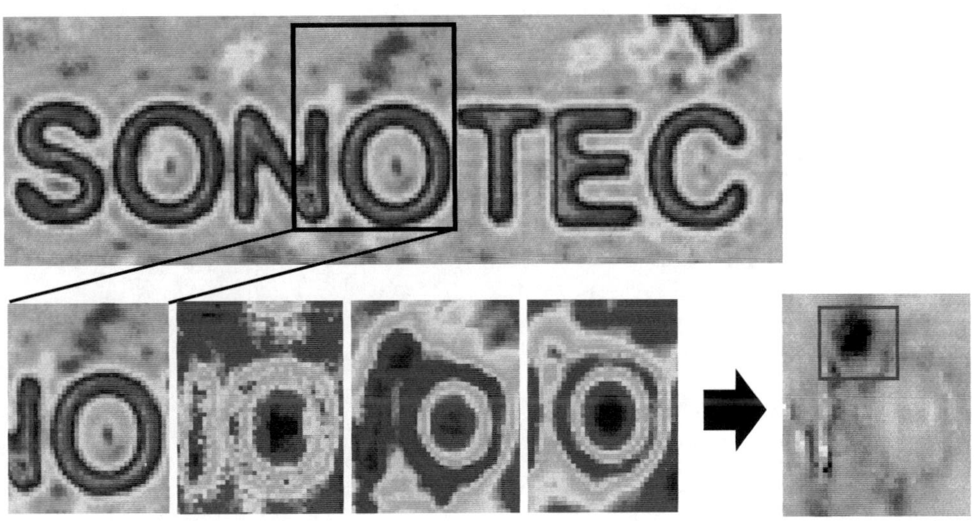

NDT in Canada 2018 | June 19-21 | Halifax, NS

SONOTEC
ULTRASONIC SOLUTIONS

Conclusion

- A wider bandwidth can be achieved with dice and fill composites
- Spectral analysis can be used as a contrast mechanism with high bandwidth probes
- The detection and sizing of delamination and flaws in composites can be improved with this contrast mechanism

NDT in Canada 2018 | June 19-21 | Halifax, NS

SONOTEC
ULTRASONIC SOLUTIONS

Ultrasound is our strength.

Headquarters Germany

SONOTEC Ultraschallsensorik Halle GmbH
Nauendorfer Str. 2
06112 Halle (Saale)
Germany
www.sonotec.eu

Application Engineer NDT
Tobias Gautzsch
t.gautzsch@sonotec.de

North American Sales and Stock Office

SONOTEC US Inc.
190 Blydenburgh Rd
Islandia, 11749 NY
USA
www.sonotecusa.com

Business Unit Director NDT
Manuel Lucas
mlucas@sonotecusa.com
+1(631)882-4132

NDT in Canada 2018 | June 19-21 | Halifax, NS

Flaw Detection Capabilities in Aerospace with Eddy Current Array Technology

Jesse Herrin

CINDE
CANADIAN INSTITUTE FOR NON-DESTRUCTIVE EVALUATION

NDT in Canada
NDTC 2018
Canada's NDT Conference
June 19 - 21
Halifax Convention Centre
Halifax, Nova Scotia

Agenda

- Common NDT Techniques for Surface Flaws
- Eddy Current Array (ECA) Testing
- ECA Inspection Benefits
- Scan Speed, Signal-to-Noise Ratio, and Sample Rate
- Friction Stir Welds
- Multi-Layer Bolt Holes
- Conclusions

Common NDT Techniques for Surface Flaws

- Liquid or Dye Penetrant Test (PT)
  - Liquid is drawn into surface openings by capillary action
  - Defects become visible under UV light or due to the dye
- Magnetic Particle Test (MT)
  - Very fine ferromagnetic particles are applied to the metal
  - Particles are drawn into flaws which indicates their presence
- Eddy Current Test (ECT)
  - Coils apply EM field into metal, flaws disrupt EC flow
  - Can find defects in ferrous and non-ferrous materials
  - Inspection results can be captured and stored

Eddy Current Array (ECA) Testing

- Multi-coil arrays take EC technology a leap ahead
- ECA probes have multiple coils in the same probe
  - positioned longitudinally, transversely, or off-axis
  - fired at coordinated times
- With an array probe, users can
  - capture more information in a single pass
  - dramatically increase speed, accuracy, and repeatability
- Inspection can now be done in a fraction of the time

ECA Inspection Benefits

- Quick, clean and portable
- Accurate, high quality inspections
- Recordable data
- Greater coverage in a single pass
- No need to store or handle chemicals
- Low total cost of ownership

Scan Speed, SNR, and SR

- ECA detectability is inversely proportional to scan speed
- Slowing down the sample rate (SR) will achieve better signal-to-noise ratio (SNR)
- A high-quality SNR ECA instrument can increase the scan speed ability while maintaining the required probability of detection (POD)
- An ideal ECA solution, consisting of instrument and probe, is one that produces the best SNR to meet these inspections needs

Scan Speed, SNR, and SR

Video Presentation in
Original e-media

NDT in Canada 2018 | June 19-21 | Halifax, NS

Friction Stir Weld Inspections

- X-ray is time consuming and access can be difficult
- PT requires handling of chemicals and surface prep
- Surface breaking cracks difficult for UT to detect and size
- Grain structure of FSW can mask flaws for UT

NDT in Canada 2018 | June 19-21 | Halifax, NS

Friction Stir Weld – ECA Solution

- ECA is an ideal technique for FSWs
- A small probe and instrument allows for fast and easy access to FSW seams
- No surface preparation or handling of chemicals required

Surface Crack Detection for FSW with ECA

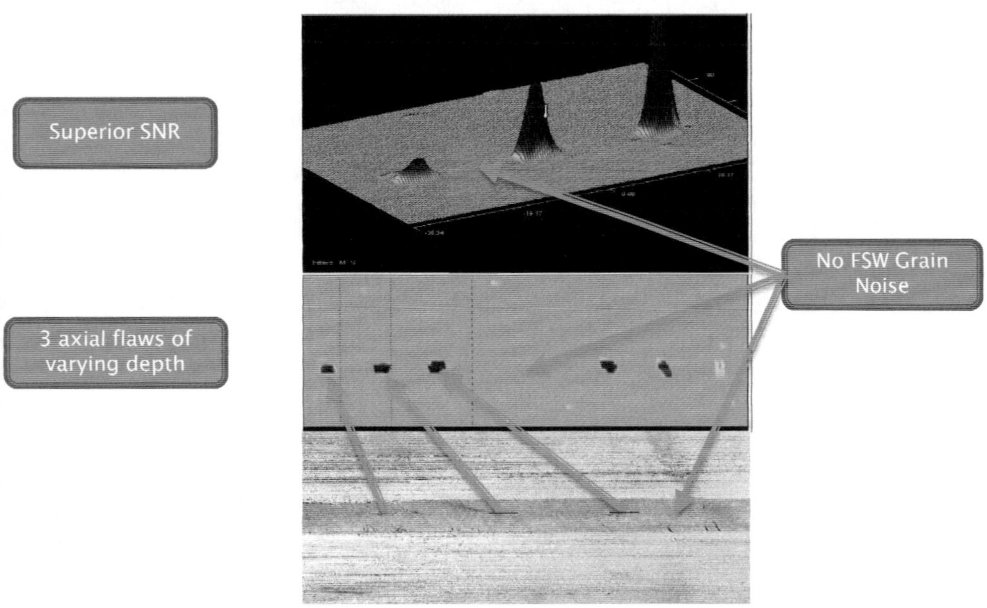

Surface Crack Characterization with ECA

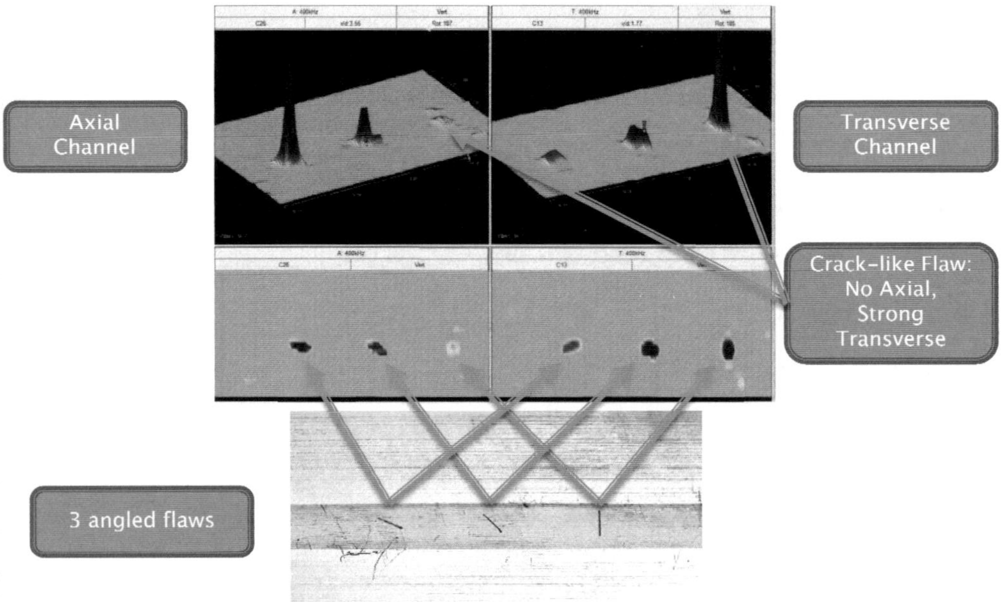

NDT in Canada 2018 | June 19-21 | Halifax, NS

Friction Stir Weld Inspections

Video Presentation in Original e-media

NDT in Canada 2018 | June 19-21 | Halifax, NS

Multi-Layer Bolt Hole Inspections

- Difficult to identify the layer in which a flaw occurs
  - Especially when flaw is close to transition between layers
- Current method requires use of a marker or tape on the probe to indicate depth

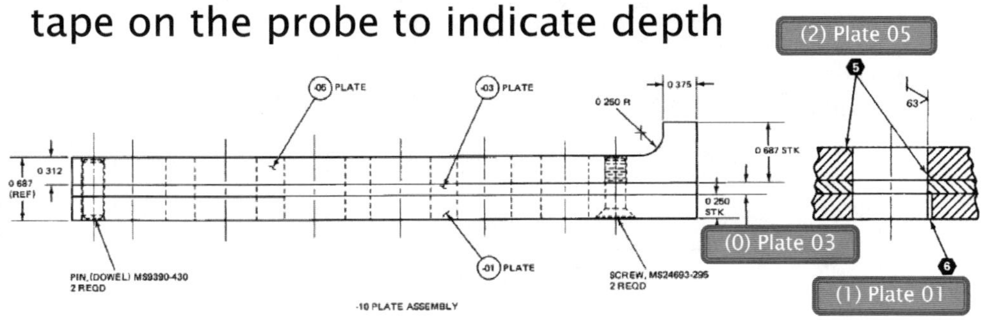

NDT in Canada 2018 | June 19-21 | Halifax, NS

ML Bolt Hole - C-Scan Solution

- C-Scan display allows the user to "see" the different layers
- Better POD and characterization of signals than standard impedance & sweep displays
- Color 2D & 3D C-Scan displays from ECA now being used for bolt hole inspections
- Superior SNR makes flaws easily identifiable

NDT in Canada 2018 | June 19-21 | Halifax, NS

Better POD for ML Bolt Hole Inspections

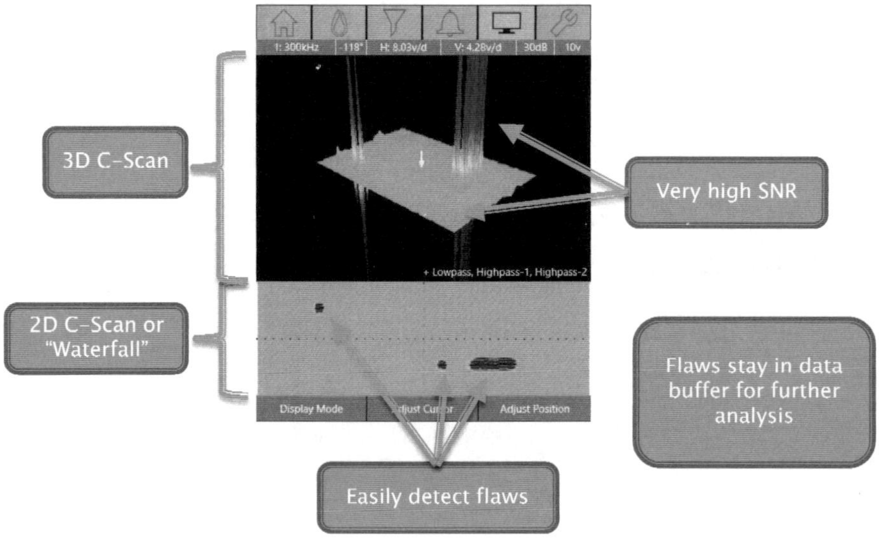

NDT in Canada 2018 | June 19-21 | Halifax, NS

Seeing Layers in ML Bolt Hole Inspections

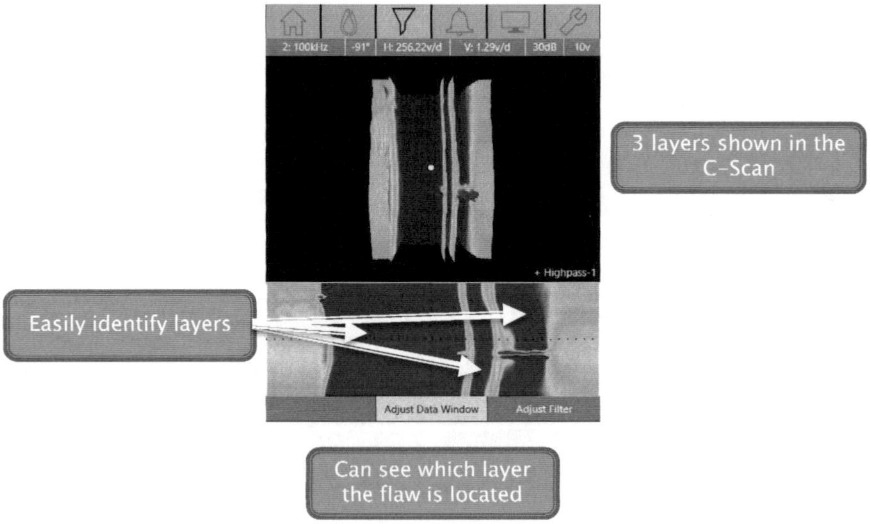

NDT in Canada 2018 | June 19-21 | Halifax, NS

Multi-Layer Bolt Hole Inspections

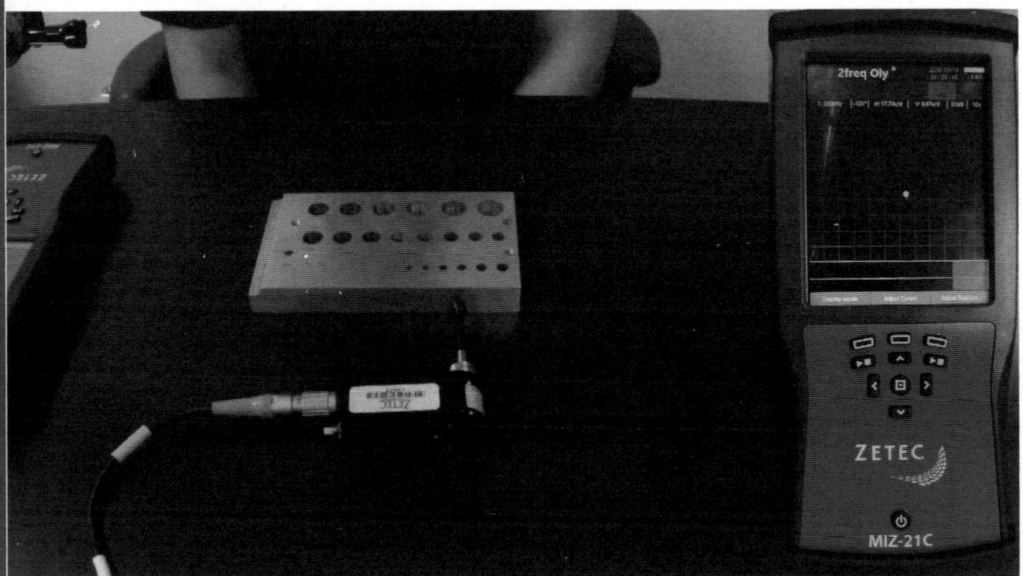

NDT in Canada 2018 | June 19-21 | Halifax, NS

ML Bolt Hole - C-Scan Conclusions

- ECA C-scans improve detection capability vs. traditional rotating scanner inspections
- Flaws can easily be assigned to a layer
- Use of an EC instrument with high SNR further improves the POD

NDT in Canada 2018 | June 19-21 | Halifax, NS

Conclusions

- A high-quality SNR ECA instrument can increase the scan speed ability while maintaining the required POD
- With higher noise level there is a greater chance of missing small flaws, therefore a high sample rate is not necessarily good; what is more desirable is a high SNR
- Surface breaking flaws can efficiently be found in FSW using an ECA probe and a handheld ECA instrument in comparison to existing techniques
- During multilayer aluminum bolt hole inspections, it can be determined in which layer flaws exist by using C-Scan and dual frequencies

Thank You

Questions?

NDT in Canada 2018 | June 19-21 | Halifax, NS

High Frequency Inspection Solution for Austenitic Stainless-Steel Welds

F. Lachance[1], P. Rioux[1], D. Giguere[1] and J. Turcotte[1]

1 Sonatest, Québec, Canada

Phone: +1 418 683 6222, e-mail: sales@sonatest.com

Abstract

Austenitic stainless steel is well known for its high mechanical properties and corrosion resistance. This material is used in the oil and gas and nuclear industry, where ultrasonic techniques represent an important proportion of the inspection in the quality control process. However, these alloys often are very attenuating and the sound behaviour of this material is anisotropic. Those characteristics are caused by the grain size and orientation, especially inside the welded volume of a component. Conventional ultrasonic approaches have limited inspection performance, while phased-array with focusing capability and beamforming offers broader coverage and better resolution. This article is going to present a quantitative comparison of the signal quality between some configurations and parameters changes on highly attenuative dissimilar weld samples. Results will show that contrary to popular beliefs; quality of the electronic equipment, focusing capability and inspection configuration, along with probe frequency, can also influence the inspection quality. It will also demonstrate that it is not always required to sacrifice sensitivity and sizing capability over SNR for that kind of inspection.

Keywords: ultrasound testing (UT), phased array (PA), stainless steel, austenitic, duplex, dissimilar weld, dual linear array (DLA), transmit receive longitudinal (TRL)

1 Introduction

Stainless-steel alloys are pretty popular for their corrosion protection to chemical processes; processes that can be found in the oil and gas, power generation industries for instance. Welded engineering structures like pipework's, heat exchangers, pressure vessels and LNG thanks commonly use those corrosion resistance alloys (CRA) because they are exposed to these chemical processes every day. Whether in operation or during their manufacturing stage, the welds of these components need to be inspected to confirm their integrity and reduce the risk of failure during operations. So far, the radiographic technology has been, and is still, a very popular and accepted NDT methodology widely approved and used around the world.

With the introduction of portable phased array instruments back in the early 2000s, the ultrasound NDT technique had a huge impact on the weld inspection by bringing an alternative to the classic radiographic NDT approach. These instruments offer state-of-the-art acoustic capabilities to fulfil codes and standards requirements to carry inspection of welded structures in lieu of radiography. Introduced a couple of years ago, some pitch and catch techniques using phased array probes have shown great results for the inspection of austenitic stainless-steel weld inspections. Nonetheless, one may ask what is the difference between these techniques; what are the limitations? This paper presents a quantitative comparison between different configurations and acoustic parameters to evaluate the quality of the solutions; hence, help identify what solution is most suitable for the inspection of stainless steel welds.

2 Challenges and Limitations

Austenitic stainless steel inspection has always been perceived as a high-end inspection as it requires more equipment and expertise compared to carbon steel inspection. The use of ultrasound NDT technique for quality control of those alloys has been hampered by the physical property of this material. Its microstructure generates beam scattering, yielding a low signal-to-noise ratio (SNR) and a severe attenuation of about 6.4 dB/inch compared to 2 dB/inch for carbon steel [I]. Stainless steel also has an anisotropic property and a velocity change between the weld and the base metal which could generate mode conversion and beam distortion. Those characteristics complicate the characterisation, sizing, and localisation of potential flaws; hence, lowering the probability of detection. It is understood that the solutions and techniques discussed in this paper also apply to other attenuative materials making weld inspection a challenge. These include dissimilar welds made of Inconel or other exotic alloys per example.

2.1 Ultrasonic NDT Past and Current Techniques

The conventional ultrasound inspection is based on the optimisation of UT parameters influencing the penetration, sizing and the SNR. For austenitic alloys, the probe frequency, damping, focal distance, ultrasonic propagation wave mode and inspection angle shall be optimised for better detection and sizing. Also, the base metal and the weld configuration might have different velocities which would refract the beam and therefore generate beam distortion. The positioning error, in some cases, could reach 25% in the distance [II]. The physical limitations and challenges are the same when using the phased array technique. On the other hand, phased array brought an adaptive of the focusing capability which helps optimise the probe aperture and its focal distance. Moreover, the 2D display of the weld volume gives a more intuitive representation of the beam distortion assessment, and the multi-angle inspection eliminates the optimisation requirement of the refracted angle towards the grain orientation. Finally, the recommended phased array technique to inspect austenitic stainless steel welds is often based on a pitch and catch low-frequency configuration reusing the conventional UT approach to optimise the SNR.

3 Method

With reference to Figure 1, the side drilled hole of a highly attenuating dissimilar weld sample has been used as a reference to evaluate the effect of ultrasonic parameters on the signal.

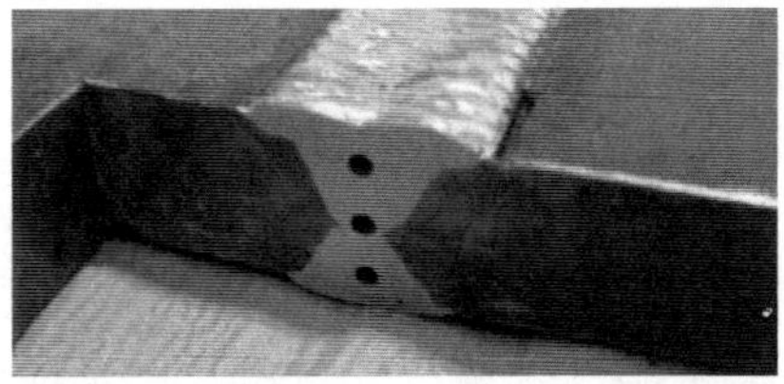

Figure 1: Weld sample used for the study

3.1 Calibration and SNR Evaluation

Since the purpose of next section is to compare the acoustic parameters, calibrating every tested configuration could have generated human measurement errors in the manipulation. Therefore, no calibration has been realised before the data acquisition, and the theoretical value of stainless steel velocity has been used. Regarding the time-corrected-gain calibration (TCG), it is recognised in the industry that it is practically impossible or very difficult to build a meaningful TCG curve on an anisotropic material. The SNR formula used everywhere in this paper (1) has been taken from the European standard EN12668-3: 3.4.3.2. The SNR measurement is one way among others to assess the quality of an inspection system.

$$SNR = \text{dB } 80\%_{Noise} - \text{dB } 80\%_{Hole} \quad (1)$$

4 PA Parameters Optimization - Tests and Results

4.1 Wave Mode Comparison

The sound propagation inside the material can is identified by the ultrasound wave mode. The longitudinal wave mode (LW) is characterised by its parallel orientation propagation and the shear wave mode (SW) by its perpendicular orientation propagation. If the shear wave mode is predominant for carbon steel inspections, the longitudinal wave, in some situations, has been proven more effective for austenitic stainless-steel inspection. Both modes have pros and cons; however, the longitudinal wave is less affected by the microstructure of the material. This behaviour has a direct impact on the SNR as demonstrated in Figure 2. In this case, using the same 5MHz 32E probe, LW mode gives almost 6 dB more of SNR compared to the SW mode.

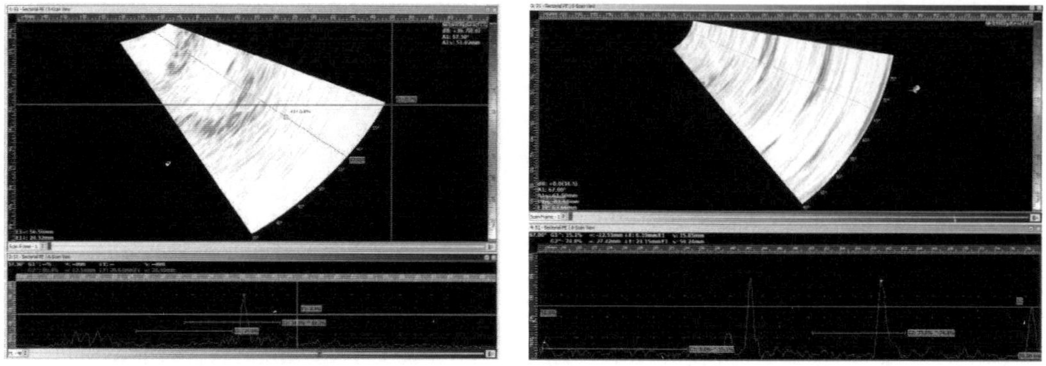

SW - SNR: 9.6 dB LW - SNR: 15.1 dB

Figure 2: Wave mode comparisons

4.2 Phased Array Active Aperture

The number of elements, hence the active aperture used for focusing and steering, shall have a great influence on the readings in terms of amplitude, resolution, and sizing. In Figure 3, a moderate improvement is observed on the 32 elements apertures vs. the 16 elements one. Generally speaking, the more elements used for focusing, the tighter the beam would be but the better the sizing and penetration capability it should produce.

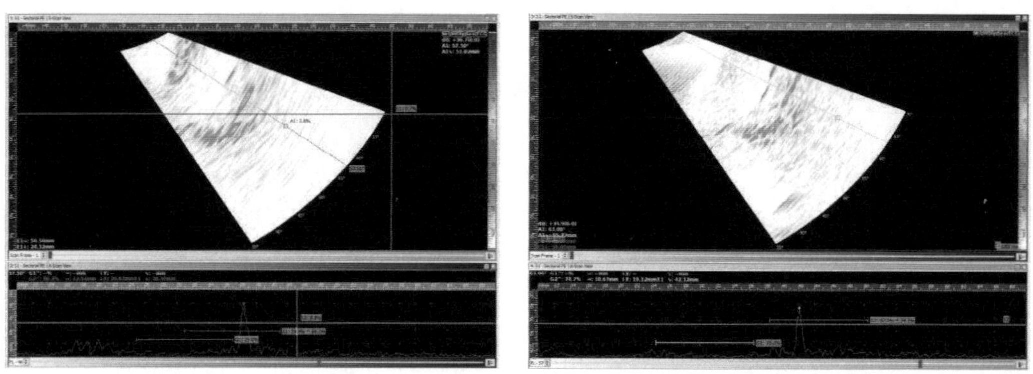

SW 16E - SNR 9.6 dB SW 32E - SNR 13.6 dB

Figure 3 Element number comparison

4.3 Phased Array Probe Frequency

4.3.1 Shear Wave

As demonstrated in point 4.1, the shear wave is the propagation mode that is the most affected by the microstructure of the material. The beam scattering is generated by the grain boundary and is independent of the inspection frequency. However, the sound attenuation is directly linked to the frequency which explains the fact that the SNR increases as the frequency decreases (Figure 4). Even if this wave mode is more affected by the microstructure of the material, it has some clear advantage compared to the longitudinal wave inspection and should be privileged when possible. It has been proven effective in the past on some austenitic stainless steel weld configurations, and it is strongly recommended for thicknesses between 2 and 10 mm [III].

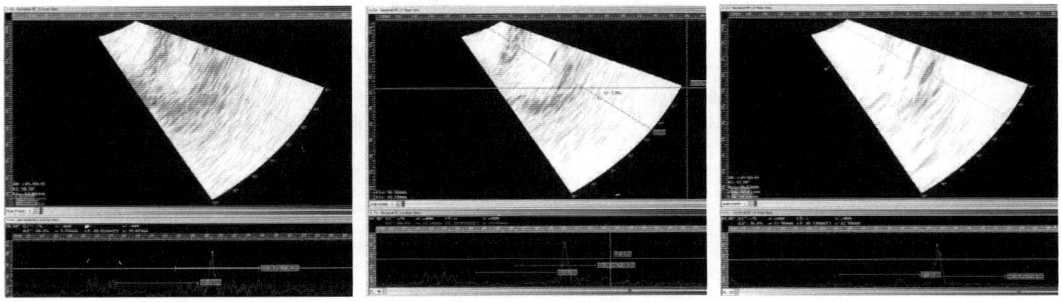

7.5 MHz - SNR 7.5 dB   5 MHz - SNR 9.6 dB   2.25 MHz – SNR 15 dB

Figure 4 Frequency comparison for shear wave

4.3.2 Longitudinal Wave

The longitudinal wave mode is less affected by the microstructure of the material. The beam scattering is smaller making it less dependent on the wave orientation with regard to the microstructure of the weld. Therefore, the effect of the frequency on the SNR is not following the same logic as for the shear wave mode. The longitudinal wavelength is approximated to twice the length compared to the shear wave; it has a direct effect on the sensitivity of the inspection. The sharpness of the beam and the sizing capability are automatically decreased for lower frequencies. Those reasons explain why we had worse results (Figure 5) using a 2.25-MHz probe compared to the 5-MHz one.

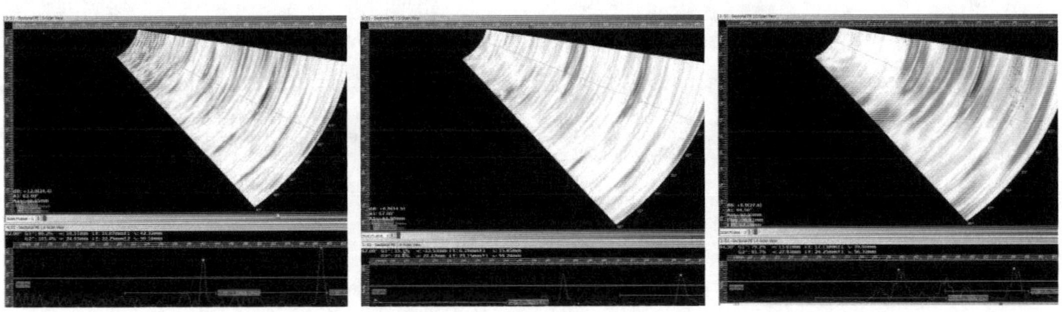

7.5 MHz - SNR 9.1 dB   5 MHz - SNR 15.1 dB   2.25 MHz – SNR 10.2 dB

Figure 5: Frequency comparison for longitudinal wave

4.4 Phased Array UT Configuration Technique

The following phased array ultrasonic configurations have been compared in order to quantify their effects on an inspection: the standard pulse-echo (electronic pulses and receives on the same elements), the pitch and catch tandem (electronic pulses and receives on different elements set within the same wedge), and the pitch and catch dual linear array DLA or transmitter-receiver-longitudinal TRL (electronic pulses and receives on different elements from isolated wedges set side by side), as represented in Figure 6.

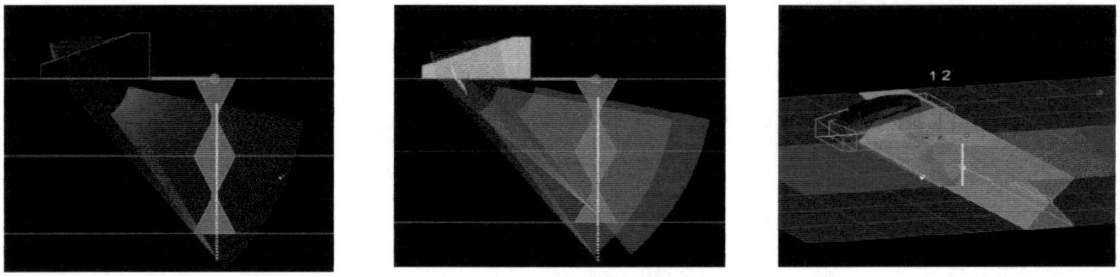

Pulse-Echo Pitch & Catch (tandem) Pitch & Catch (DLA)

Figure 6: UT configurations

With today's 32 PR (32 elements pulsers & receivers) electronic architectures, the pitch and catch phased array technique is recognised to greatly increase the SNR of an inspection by decreasing the signal noise. Indeed, by receiving on different elements than those who emitted or by receiving on elements connected to an isolated wedge, the induced sources of noise are reduced. Results shown in figure 7 support this; SNR is improved between PE and tandem, and the improvement is even more blatant between tandem and DLA. This being observed, the DLA configuration also has a smaller footprint than the other two configurations because there is no need for damping material to be integrated into the wedge design. The smaller footprint is an advantage for the DLA because it reduces the probe distance to the weld, improving the SNR at the same time.

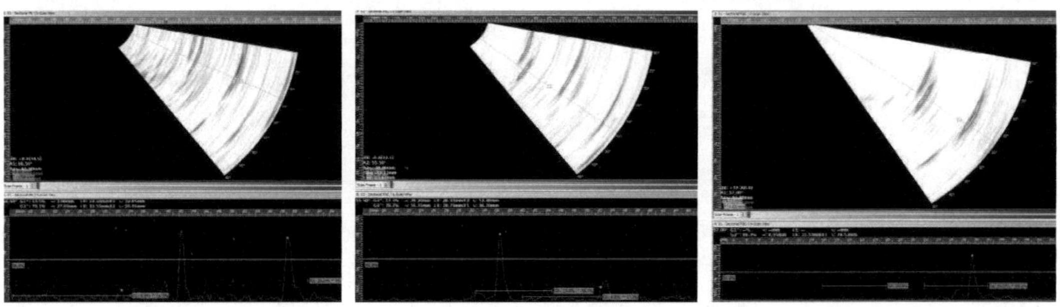

Pulse-Echo - SNR 15.1 dB Pitch & Catch Tandem SNR 23.7 dB Pitch & Catch DLA SNR 33.3 dB

Figure 7: UT configuration comparison

4.5 Phased Array Electronic Equipment

The instrument used obviously plays a key role in the overall results. The electronic circuit design does have a direct impact on the signal quality. Though good SNR instruments might not be critical for low attenuation carbon steel inspections, it can be critical for certain stainless steel inspections. Results in Figure 8 show clearly that for the same configuration parameters, the Sonatest Veo+ electronic circuitry significantly improves the signal quality. On highly attenuated material like austenitic stainless steel, it could bring a real valuable advantage.

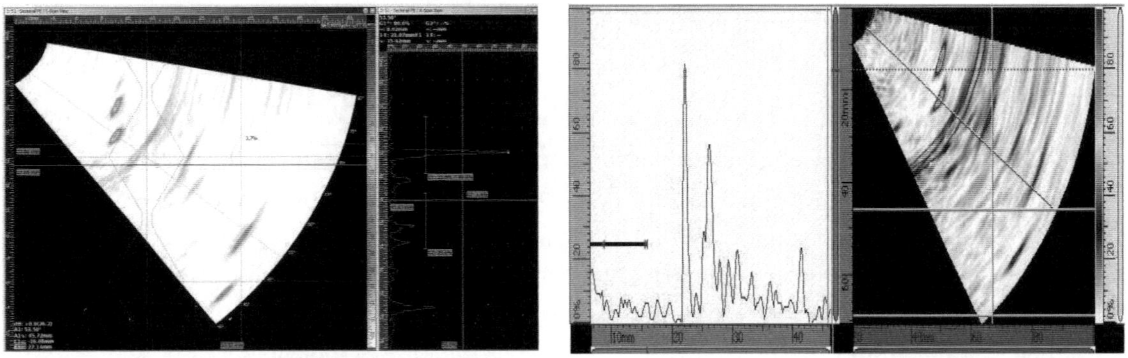

Sonatest Veo+, 50-V pulser
5 MHZ, PE, 32E, LW
SNR: 14.1 dB

Other phased array equipment, 80-V pulser
5 MHz, PE, 32E, LW,
SNR: 3 dB

Figure 8: Electronic equipment comparison

5 Phased Array DLA Solution - Technique, Tests & Results

In the previous sections, the relative effects of some parameters on the SNR have been demonstrated. From the good results associated to the pitch and catch dual linear array (DLA) acoustic configuration observed in the preceding sections, Sonatest had developed a DLA solution dedicated to austenitic stainless steel, dissimilar and other attenuative alloys weld inspection. The results in this section have been generated using the Sonatest veo+ 32:128PR phased array system, the series of detachable active array head (DAAH) phased array probes and a new line of wedges specially designed to fit the requirements of the pitch and catch DLA and TRL technique (Figure 9).

Figure 9: Sonatest DAAH DLA probes and wedges for TRL technique

5.1 Phased Array DLA - TRL Technique

The phased array DLA solution for an austenitic stainless steel is similar to the recognised TRL conventional UT technique as it is based on a pitch and catch configuration using longitudinal wave (LW) mode. In practice, the phased array DLA LW mode is used without half skip because the back wall echo would create converted SW modes and some noise coming with it. The fact that the phased array LW mode can steer at very high angles will also help to inspect not only the volume of the weld but also its surface where possible.

In the TRL technique using DLA probes, the wedge design is key. The Figure 10 shows that the cutting angle of the wedge is optimised for LW propagation and there is a roof angle on each wedge, creating a natural focal point. The combination of those two angles will create an effective acoustic area into the weld and the phased array beamforming will ensure a coverage for different angles. During the preparation of the inspection procedure or scan plan, the effective acoustic area and phased array angles have to match the weld thickness and geometry so the transmitter beams cross the receiver ones properly in weld zone. Finally, the wedge needs to be isolated in the middle to get benefits from the low noise pitch and catch configuration.

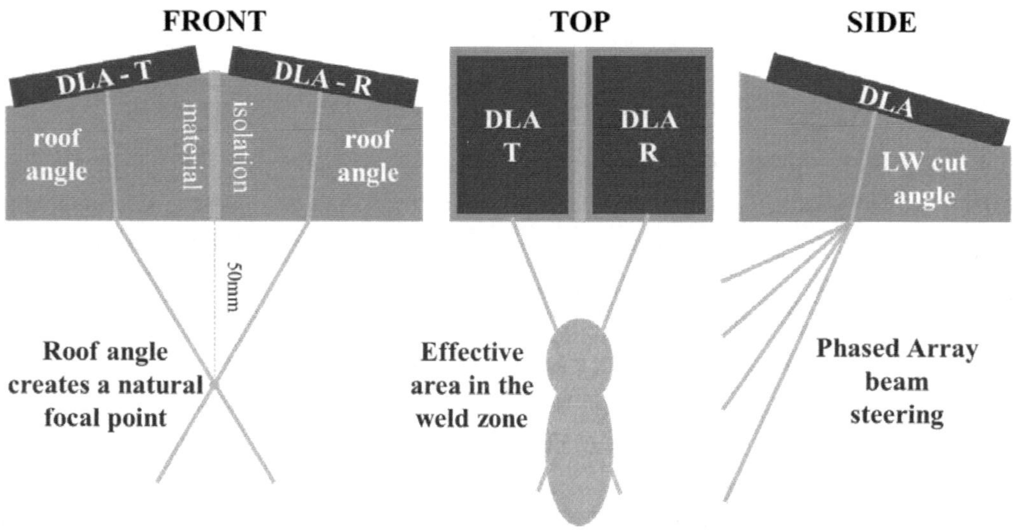

Figure 10: Phased Array DLA – TRL technique, natural focal point and effective area schematic

5.2 Phased Array DLA Tests & Results

Using the same dissimilar weld sample as in the previous sections, we compared three frequencies of DLA probe configurations. Normally, we would expect a higher SNR using a lower frequency, but surprisingly the 5 MHz offers a better result than the lower 2.25 MHz, refer to Figure 11. Because of the wide variety of alloys, tests using different probe frequencies is important to identify the best solution for inspection since the higher the probe frequency, the sharper the sizing resolution will be.

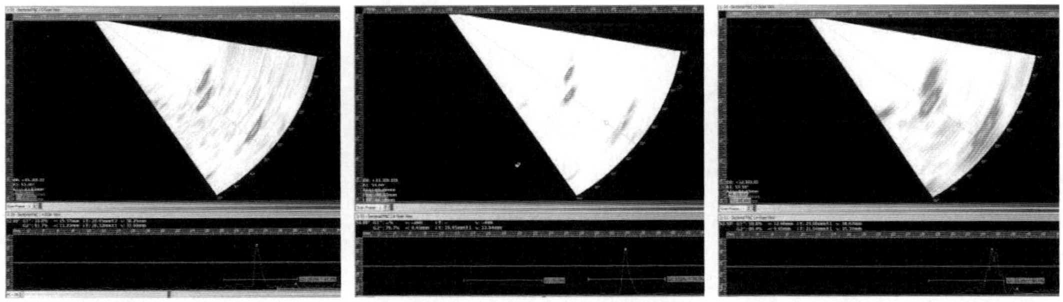

7.5 MHz DLA - SNR 17 dB   5 MHz DLA - SNR 30.6 dB   2.25 MHz DLA - SNR 19.2 dB

Figure 11: DLA solution frequency comparison

Figure 12 shows another sample where the comparison between 5MHz SW and DLA-TRL inspection in the half-skip. The significant 16dB SRN improvement shows the superior performance of the DLA solution over standard SW over a real flaw.

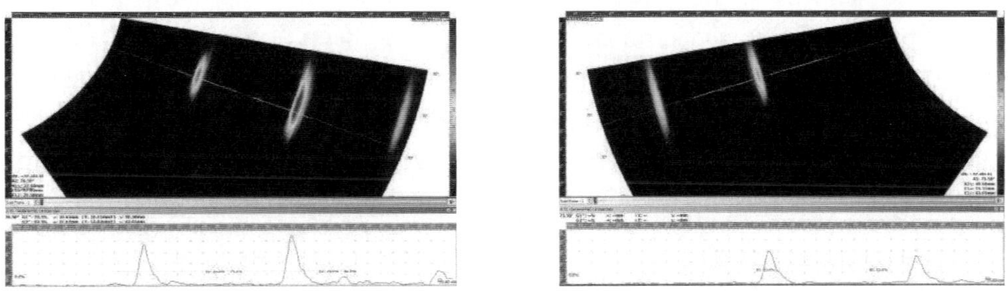

DLA LW in the half-skip – SNR 26 dB

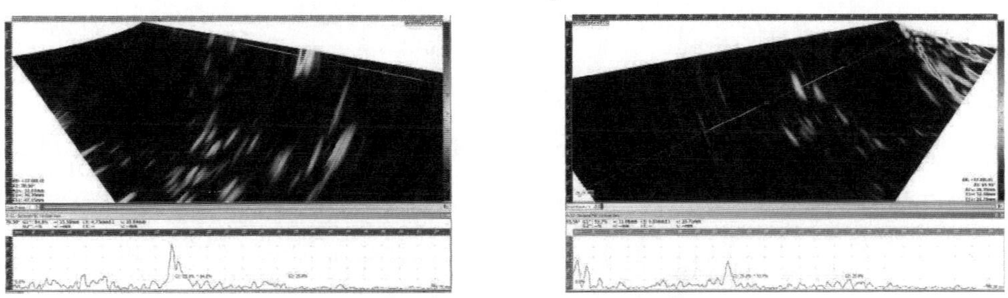

PE SW in the half-skip – SNR 10 dB

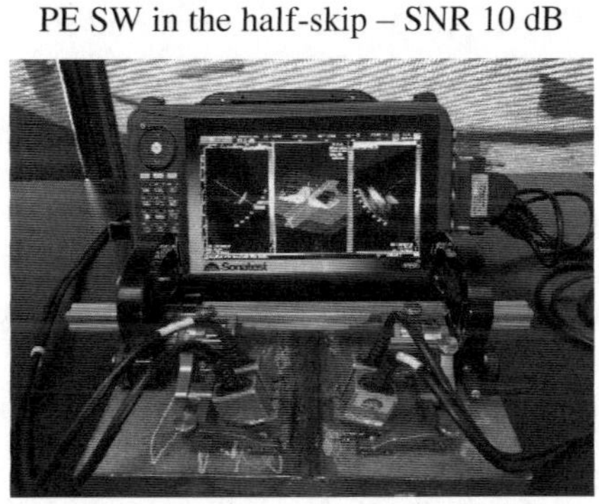

Figure 12: Two-sided inspection comparison on first leg (half-skip) SW vs. DLA-TRL

6 Conclusion

Because of the core variations in the grain structure (coarse grain) of the base material, the solidification point of the welded material and associated residual stresses combined with load induced stresses in the weld area,[IV] it is impossible to define a perfect NDT solution for all stainless steel and other exotic alloys. This paper presented different ultrasonic approaches and compared the key parameters that could affect the signal quality, and ultimately the inspection results. Results show that other parameters like the quality of the electronic equipment, the focusing capability, and the inspection configuration can also influence the inspection results. Of course, the frequency is important and shall be taken into consideration, but it has also been demonstrated that it is not always required to sacrifice sensitivity and sizing capability over SNR for that kind of inspection.

For the challenging inspections like highly attenuative austenitic stainless steel welds, the tested methodology using the Sonatest DAAH phased array DLA probes, TRL technique and veo+ instrument showed promising results. It has been noted that a standard shear wave (or LW if possible) PE technique shall be prioritised when possible because of the simplicity of application. The calibration of the PE technique, for example, is much easier than the calibration using the phased array TRL technique. The team intends to pursue the investigation on this inspection technique and propose some guidelines for the calibration.

References

[I] J.D. Lavancer , ULTRASONIC TESTING OF STEEL CASTINGS, by steel founder's society of America ,United states of America, Chap. "Ultrasonic attenuation Carbon, Low alloy and austenetic steels", 1976, pp 26

[II] R.J Hudgell and B.S. Gray, The ultrasonic inspection of austenitic materials-State of the art report, Risley Nuclear Power Development Laboratories, May 1985, pp 8

[III] JUVA,A and Lieto, A., The ultrasonic examination of thin austenitic stainless steel butt welds British Journal of NDT, July 1980

[IV] Bulavinov, Kröning and Walte, Ultrasonic Inspection of Austenitic and Dissimilar Welds, 4th Conferencia Panamericana de END Buenos Aires – Octubre 2007

Sidewall Embedded Electromagnetic Measurement of Matte level in Pyrometallurgical Furnaces

HATCH

Aroba Saleem[1], P. Ross Underhill[1], David Chataway[2], Terry Gerritsen[2], Afshin Sadri[2] and Thomas Krause[1]

[1]*Department of Physics and Space Science, Royal Military College of Canada*
[2]*Hatch, Mississauga, Ontario, Canada*

CINDE CANADIAN INSTITUTE FOR NON-DESTRUCTIVE EVALUATION

NDT in Canada NDTiC 2018 Canada's NDT Conference | June 19 - 21 Halifax Convention Centre Halifax, Nova Scotia

2

Outline

- Overview
- Eddy current testing
- Materials and methodology
- Results
- Summary

NDT in Canada 2018 | June 19-21 | Halifax, NS

HATCH

+ Introduction

HATCH

Introduction

- Determine when to tap excess matte
- Better understanding of melt times
- Monitor refractory wall condition

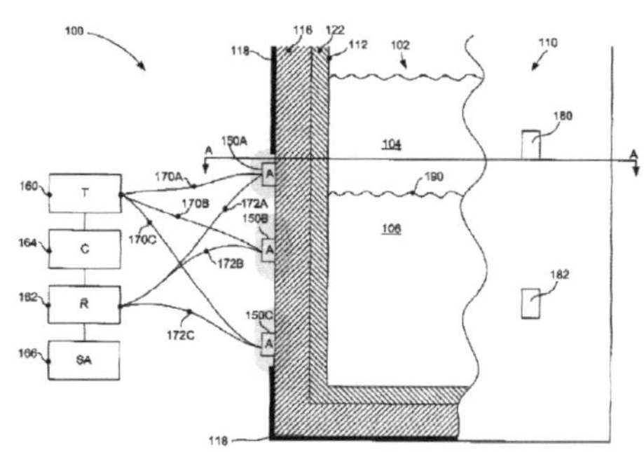

Figure: Hatch Patent

Common practice

- Sounding Bar
  - Low accuracy
    - Human interpretation
  - Not continuous
  - Safety concerns

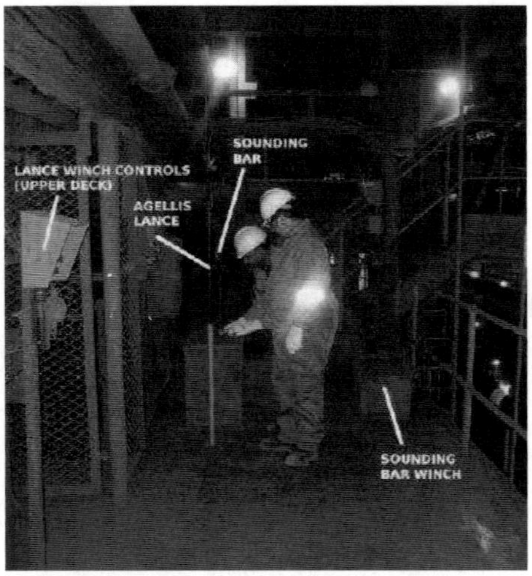

Figure: Sounding bar at Vale

Common practice

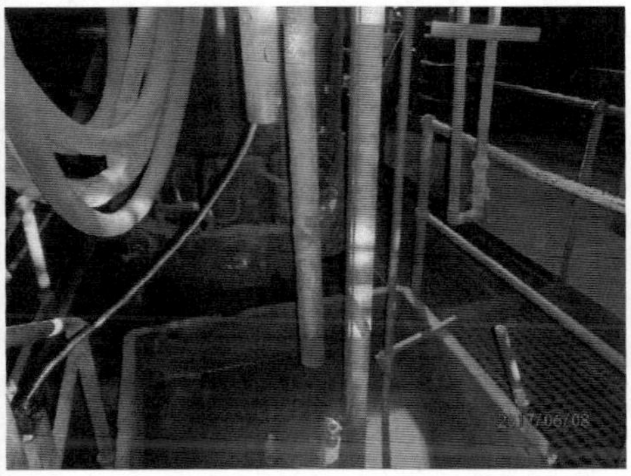

Figure: Sounding bar at NKS

Figure: Sounding bar measurement at PFK

Technologies Tested

- Hatch:
  - Electrochemical oxygen
    - Lonmin, Anglo Platinum (2004)
  - Electric potential (Stinger)
    - Lonmin (2016)
- Agellis Group AB:
  - Eddy current probe
    - Lonmin (2011)
    - Vale Sudbury, Onca Puma (2013)

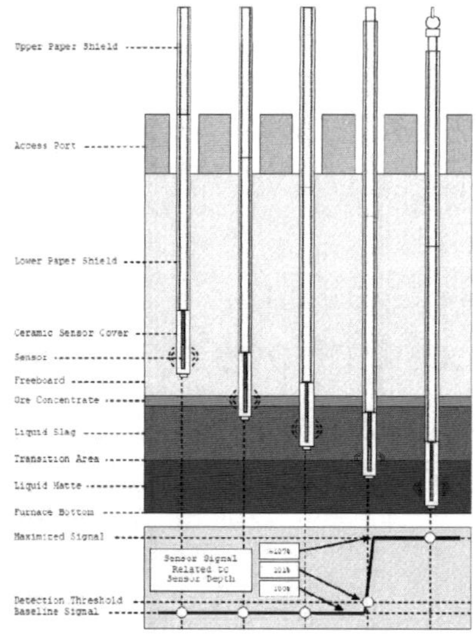

Figure: Agellis Probe

Eddy current NDT

- Matte/slag = conductive/non-conductive interface

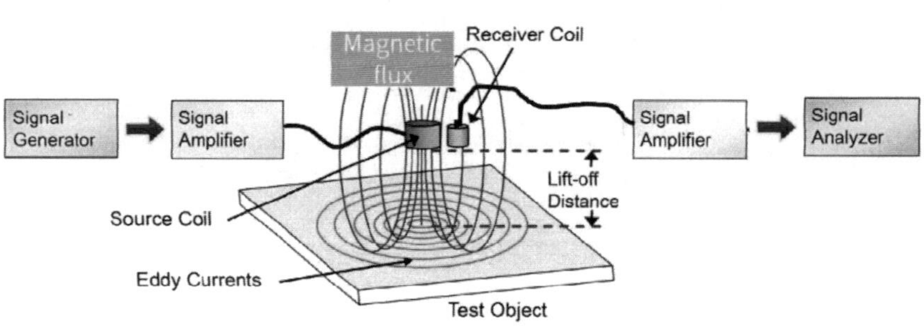

Figure: Eddy Current System

Eddy current NDT

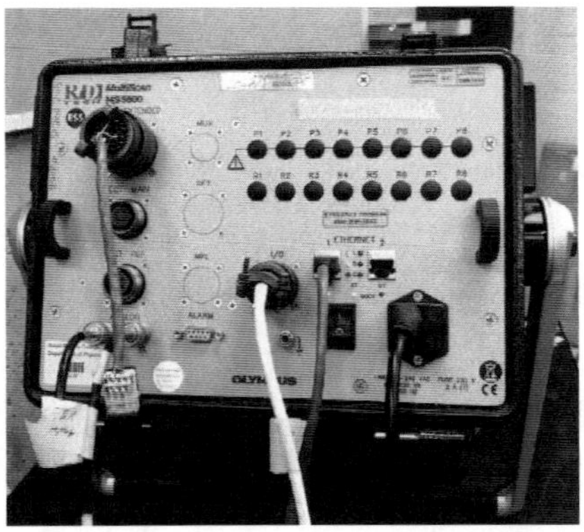

Figure: RMC eddy current application

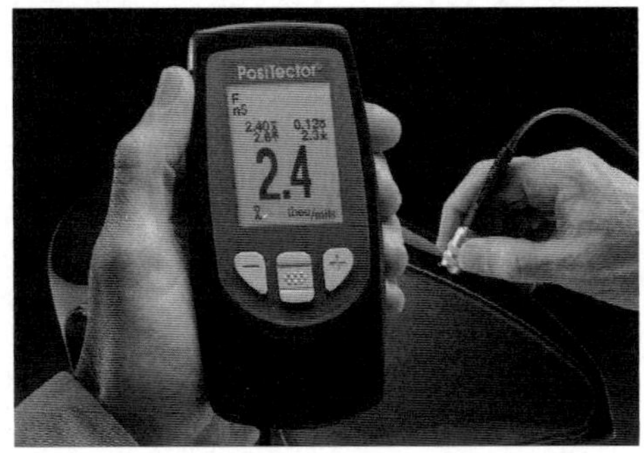

Figure: Eddy current coating thickness testing

Challenges

- Sensor location
- Refractory erosion
- High temperature
- Remote detection
- Metal shielding
- Background noise

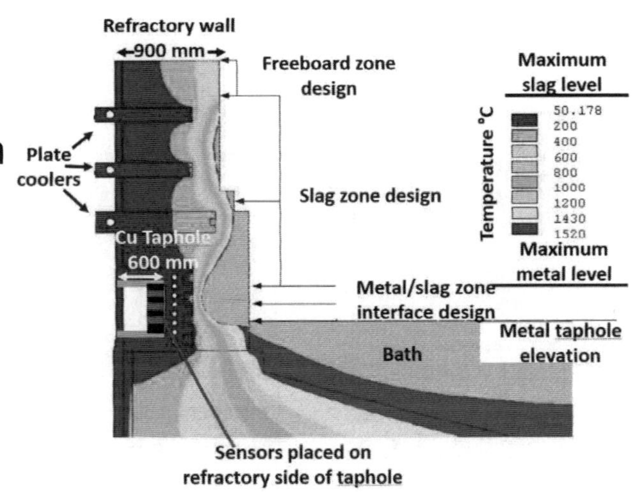

Figure: EAF Thermal Profile

Approach

Model and design system for:

1. No cooling elements
   - Impregnated refractory
2. Plate coolers
3. Waffle coolers
   - Taphole

Proposed System: Plate Coolers

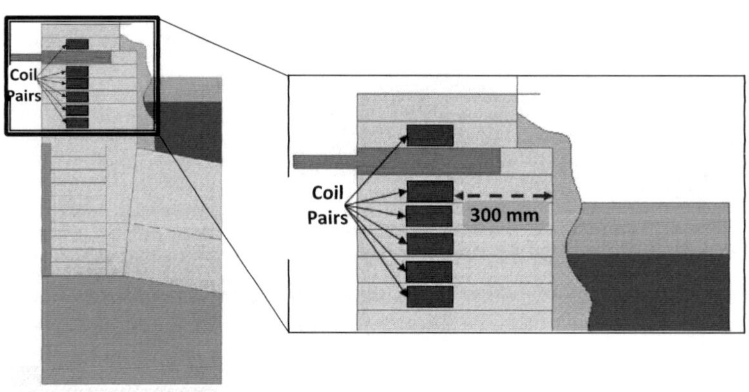

Figure: Proposed Plate Cooler Setup

Proposed System: Tapblock

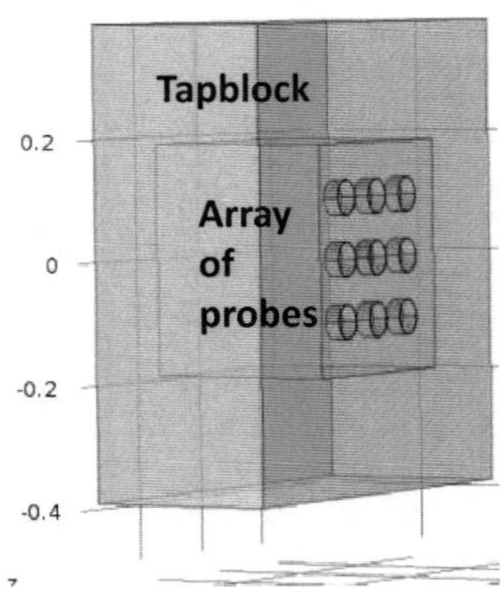

Figure: Modelled Taphole Setup

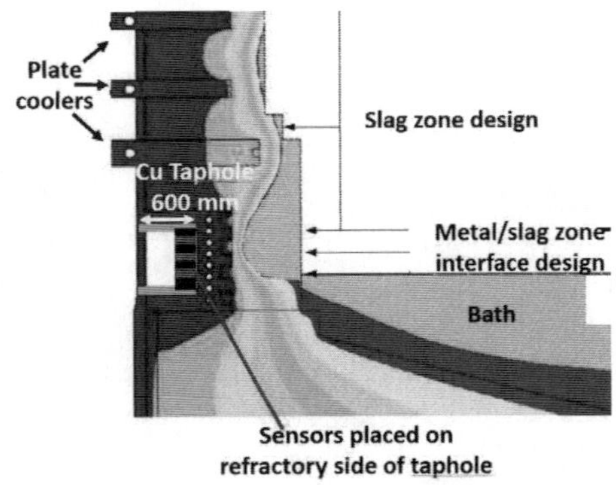

Figure: Proposed Taphole Setup

Tapblock design

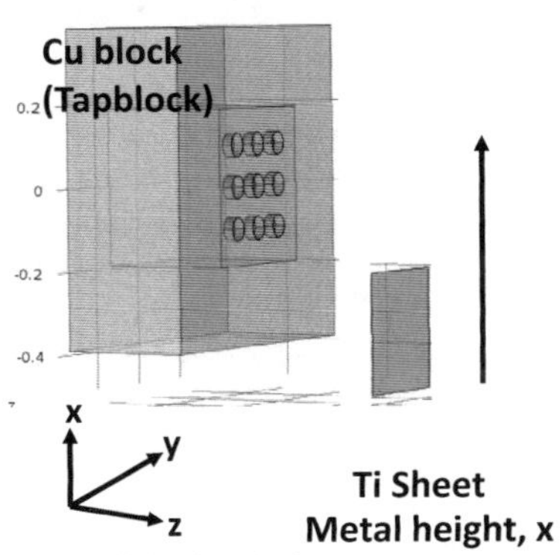

Figure: Modelled taphole geometry

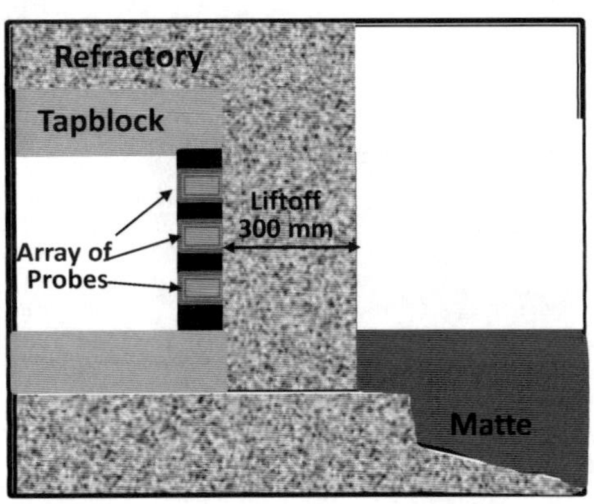

Figure: Rising matte level in a furnace

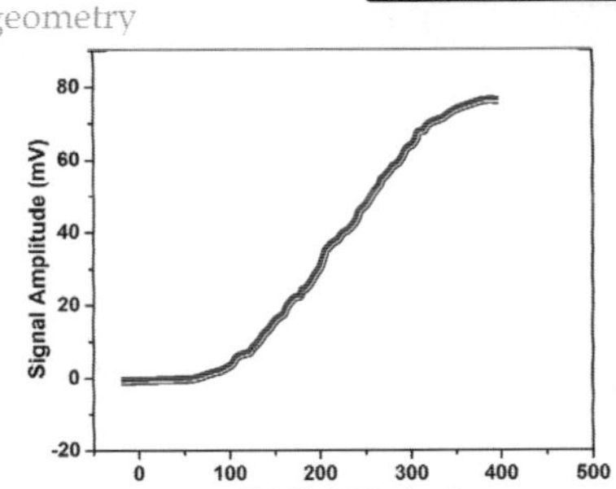

Figure: Output signal

Tapblock design

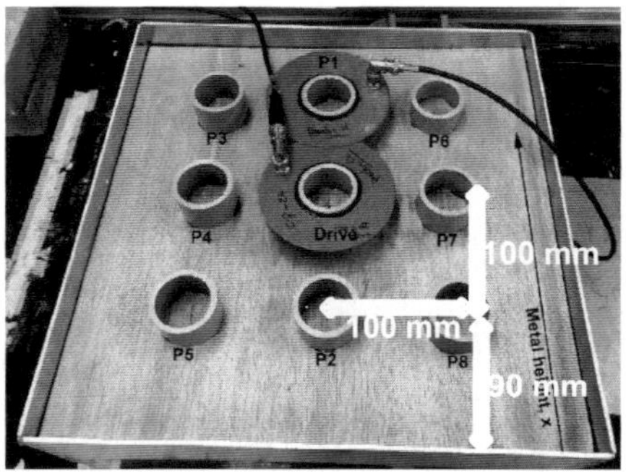

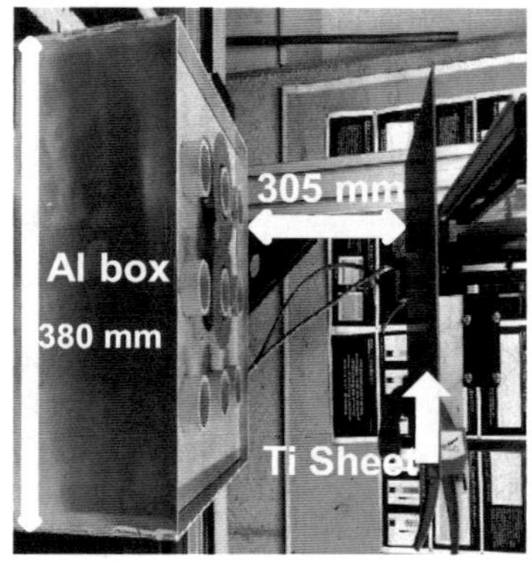

Results - Tapblock

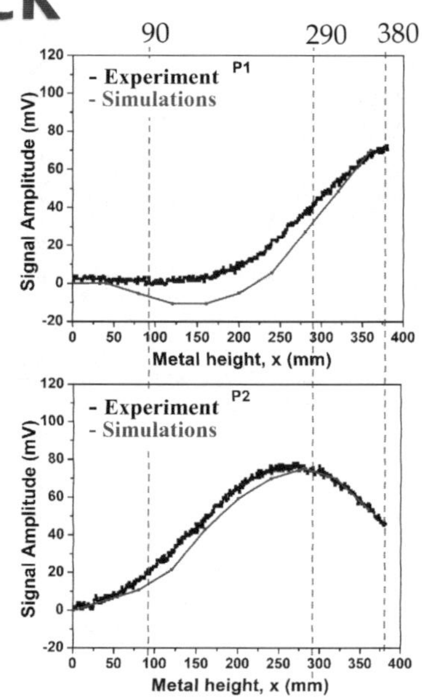

- Resistivity of Ti = 50 μΩ.cm , Cu = 1.72 μΩ.cm
- Resistivity of slag = ~ 4 μΩ.cm × 10^6 μΩ.cm
- 16 % signal increase with Cu/Al sheet, compared to Ti sheet

Projection to Latent Structures (PLS) model

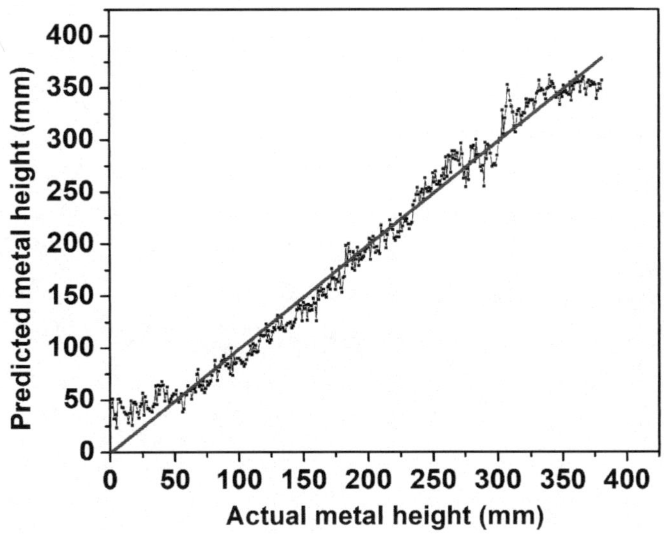

Effect of Liftoff

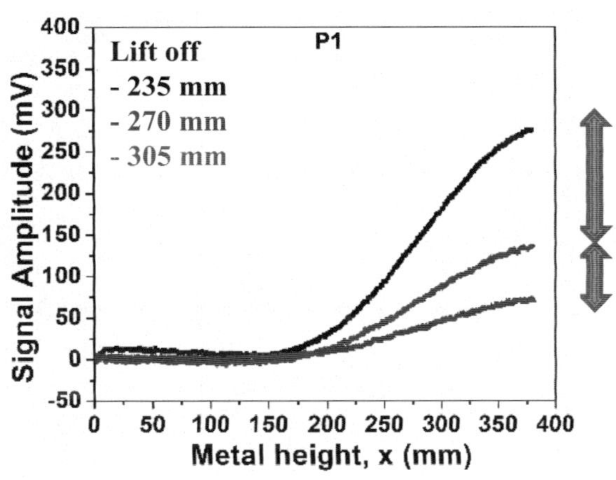

Tapblock depth

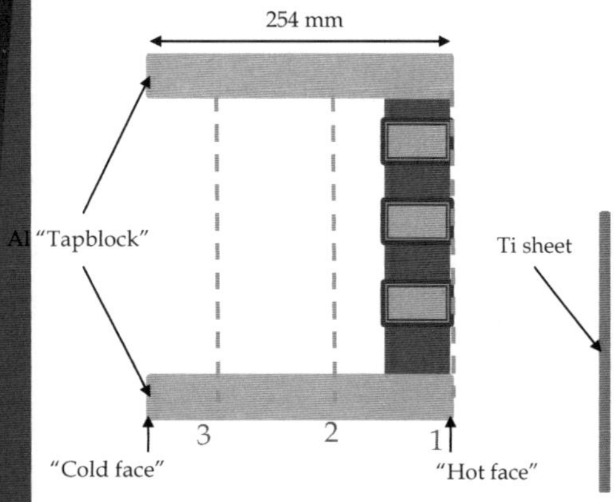

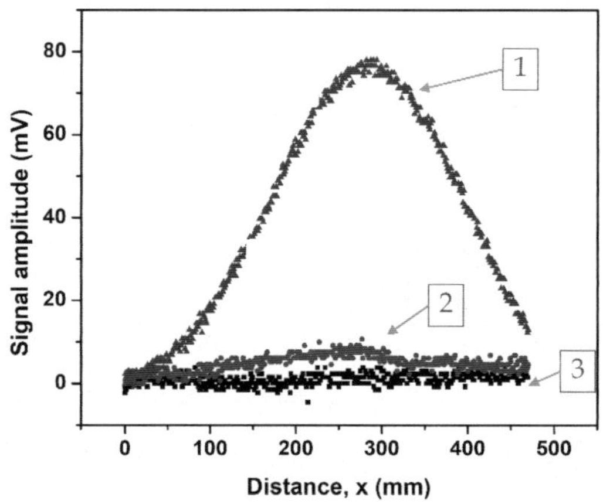

Brick Impregnation

- Impregnated brick between sensor and metal
  - Skew brick from Vale FF2

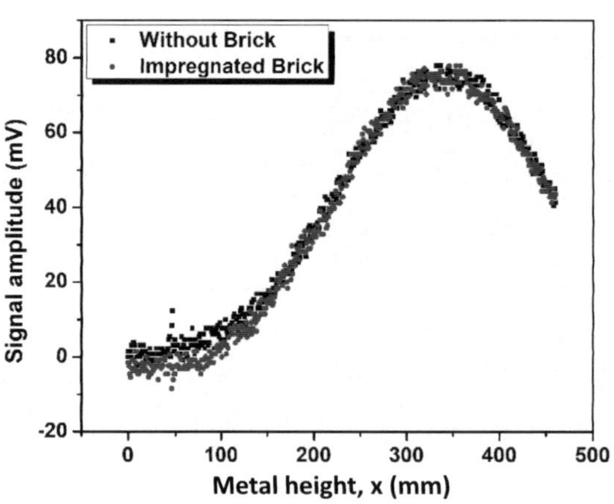

Plate cooler design

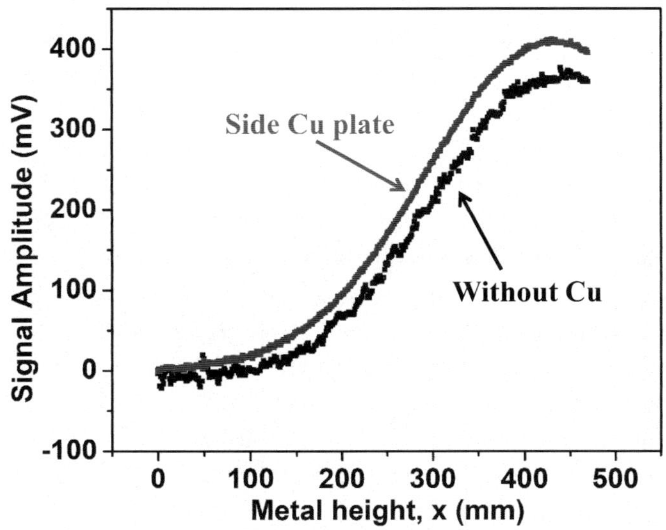

Summary

- Matte level determined at ~300 mm liftoff
  - High resolution (~5 mm)
- Signal decreases exponentially with liftoff
- Impregnated bricks have negligible effect
- Copper plates do not change the signal significantly

Summary

- Matte level sensitivity demonstrated in lab for:

1. No cooling elements
2. Plate coolers
3. Waffle coolers – at tapblock hot face

Acknowledgements

NSERC
CRSNG

Ontario Centres of Excellence
Where Next Happens

NDT in Canada 2018 | June 19-21 | Halifax, NS

Thank you

NDT in Canada 2018 | June 19-21 | Halifax, NS

Probability of Detection of Computer Analysed Pulsed Eddy Current Data

P.R. Underhill, C. Uemura, D. Butt and T.W. Krause

Department of Physics, Royal Military College of Canada, Kingston, ON, K7K 7B4

CINDE CANADIAN INSTITUTE FOR NON-DESTRUCTIVE EVALUATION

NDT in Canada
NDTiC 2018
Canada's NDT Conference
June 19 - 21
Halifax Convention Centre
Halifax, Nova Scotia

Outline

- Application
- Approach
- Intro to Hit/Miss POD
- Results
- Summary

Application

Aurora (P-3 Orion)

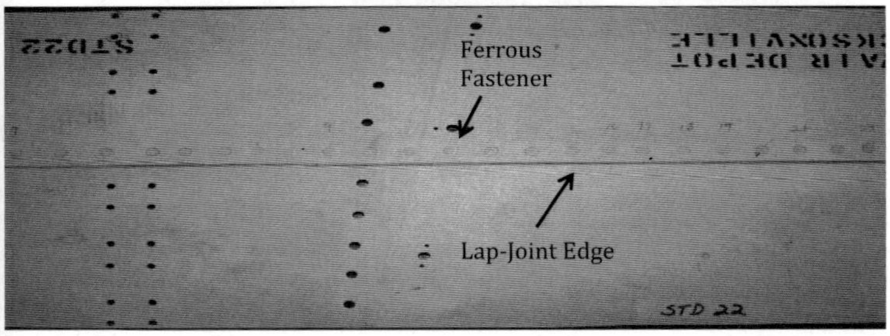

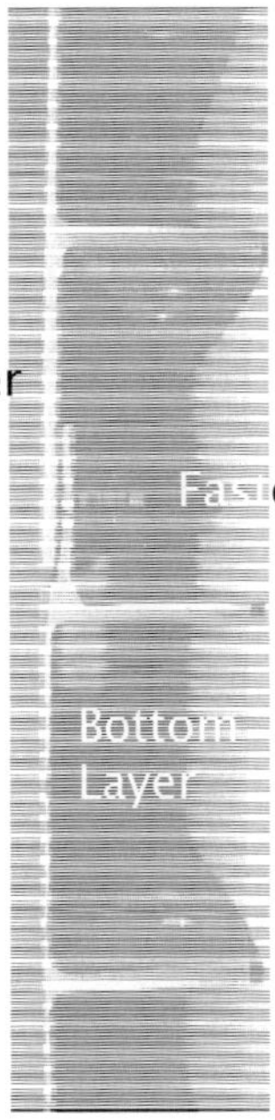

Objective

- Develop a rapid scanning technique to identify fasteners with second (and first) layer cracks without fastener removal
  - Reduce down time/cost
  - Eliminate possible damage due to fastener removal
- Use bolt hole eddy current as a confirmatory technique (sizing)

Pulsed Eddy Current Detection of Cracks

- Specially designed probe uses pulsed rather than conventional eddy current
- Generates data that is reduced to a small number of scores (5) using PCA that totally encapsulate the information

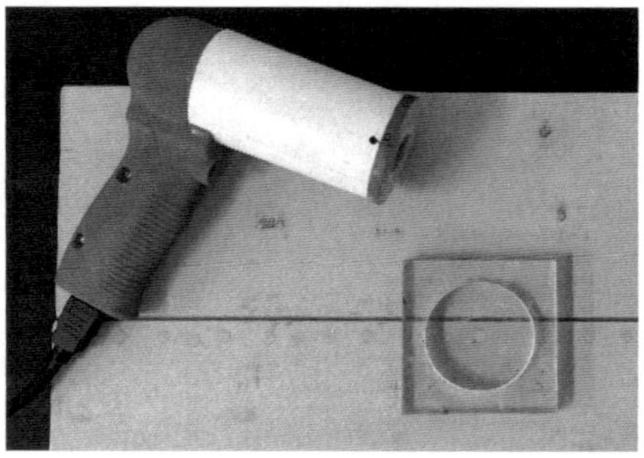

NDT in Canada 2018 | June 19-21 | Halifax, NS

Cluster Analysis

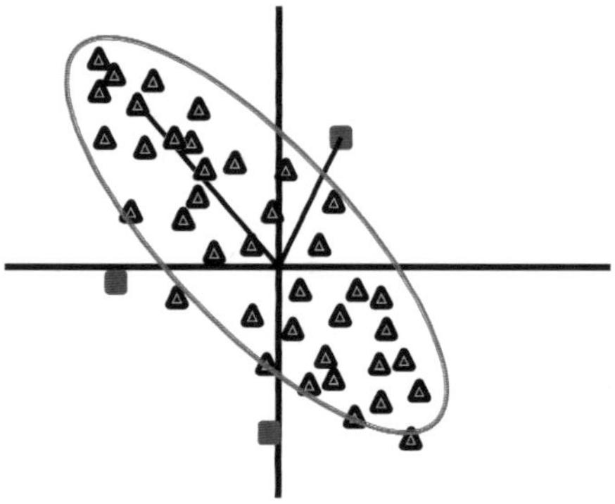

Mahalanobis Distance

NDT in Canada 2018 | June 19-21 | Halifax, NS

Characteristics

- Blind to the operator
- Treats the data as a group rather than fastener by fastener
- Uses robust statistics to overcome the blind inclusion of defects
- Previous work showed excellent ability to detect cracks
- Next step was to establish a Probability of Detection (POD)

Generate a POD

Sample Size Distribution

Angular

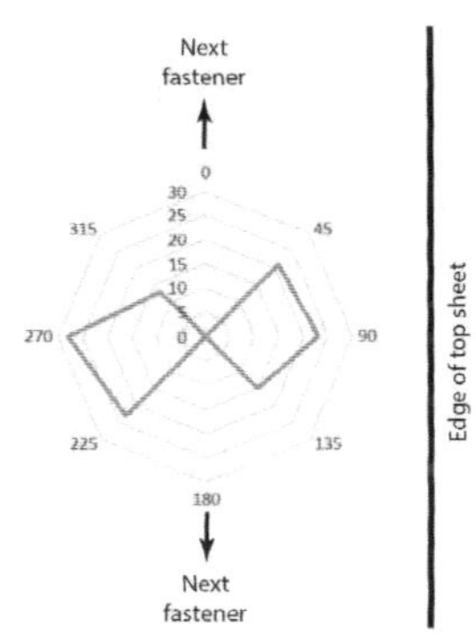

Size

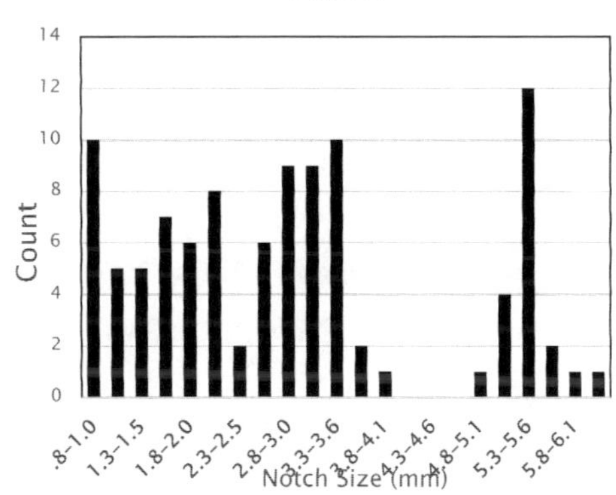

Method

- For each sample (of 7)
  - take and process data (10 passes)
  - Use bootstrap method to generate data
  - Do robust statistics (modified smallest half volume)
  - Use a pre-set cut-off to determine hits and misses
  - Repeat so that each notched fastener appears approx. 100 times
- Combine data into one large matrix
  - Approx (101 rows (fasteners) by 83 columns (trials)
- Use MIL HBK 1823 software to do POD
  - a_{50}, a_{90}, $a_{90/95}$

NDT in Canada 2018 | June 19-21 | Halifax, NS

Hit/Miss POD

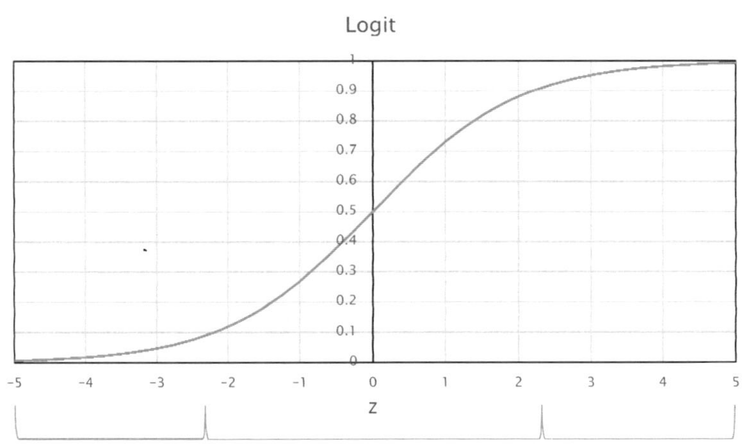

$$Z = \frac{x - \mu}{\sigma} = ln\left(\frac{p}{1 - p}\right)$$

NDT in Canada 2018 | June 19-21 | Halifax, NS

Initial Results

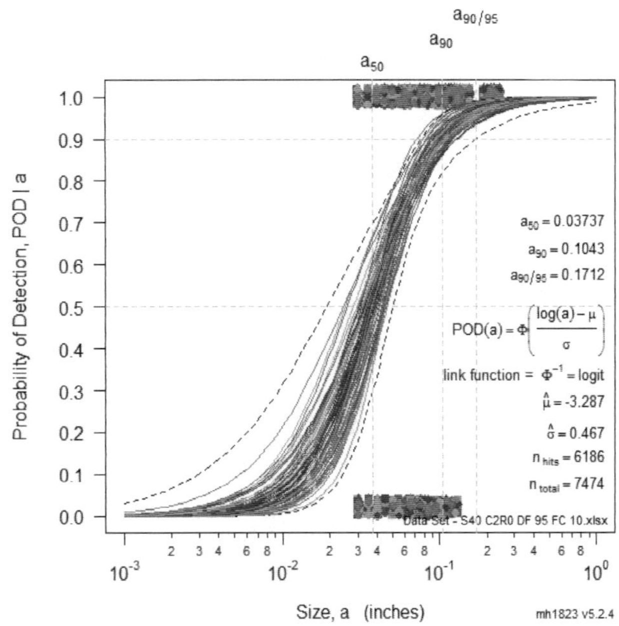

NDT in Canada 2018 | June 19-21 | Halifax, NS

Appropriateness of Fit

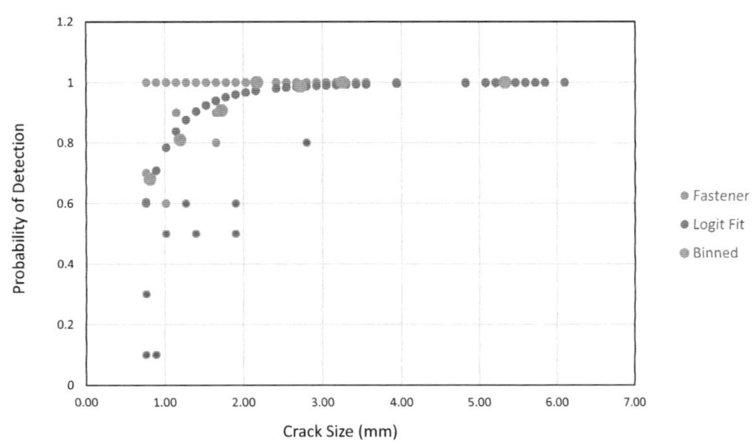

NDT in Canada 2018 | June 19-21 | Halifax, NS

Operator to Operator Comparison

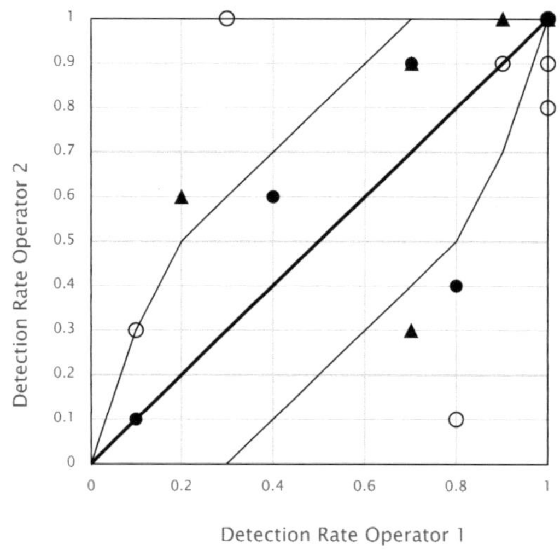

- 3 worst samples
- 45 fasteners
- 31 @ (1,1)

NDT in Canada 2018 | June 19-21 | Halifax, NS

Empirical (marginal) $a_{90/95}$

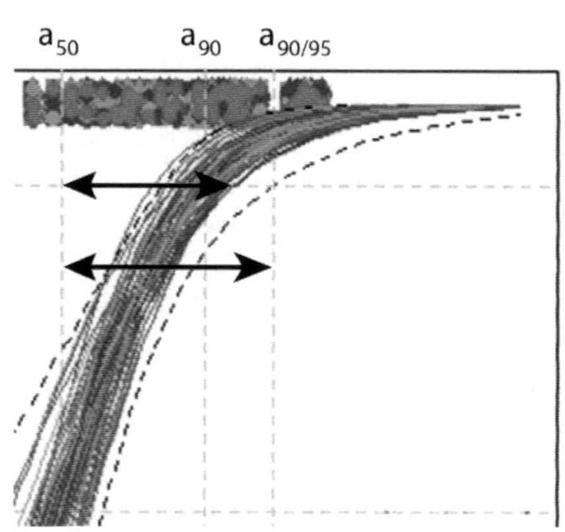

NDT in Canada 2018 | June 19-21 | Halifax, NS

Empirical (marginal) $a_{90/95}$

		Number of Notches in Simulation			
		C1	C2	C3	C4
False Call Rate	Data Fraction	$a_{90/95}$ (mm/inch) – Using ImageJ			
5% False Call	95%	2.72 / 0.107	2.74 / 0.108	3.00 / 0.118	3.10 / 0.122
	97.5%	2.60 / 0.102	2.67 / 0.105	2.90 / 0.114	2.98 / 0.117
	100%	2.43 / 0.095	2.75 / 0.108	3.24 / 0.128	3.53 / 0.139
10% False Call	95%	2.22 / 0.088	2.33 / 0.092	2.35 / 0.093	2.66 / 0.105
	97.5%	2.07 / 0.082	2.26 / 0.089	2.33 / 0.092	2.29 / 0.090
	100%	2.04 / 0.080	2.28 / 0.090	2.59 / 0.102	2.90 / 0.114
15% False Call	95%	1.93 / 0.076	2.05 / 0.081	2.10 / 0.083	2.11 / 0.083
	97.5%	1.91 / 0.075	2.02 / 0.080	1.96 / 0.077	2.08 / 0.082
	100%	1.71 / 0.067	1.93 / 0.076	2.28 / 0.090	2.56 / 0.101

Summary

- More than just size affects the detection rate
- Automated robust analysis gives a marginal $a_{90/95}$ (80 fasteners/15% FC) 2 mm/0.08"
- Have nearly 50% detection down to 0.75 mm/0.03"
- Detection rate gets worse with smaller sample sizes (40 fasteners/15% FC) $a_{90/95}$ = 2.57 mm/0.10"
- Large statistical data set makes it possible to separate out some extra effects (operator)
- Statistical approach makes it possible to generate "Quality" statistics that indicate if the inspection was carried out properly and the sample conforms to the low defect assumption

Questions

NDT in Canada 2018 | June 19-21 | Halifax, NS

Cumulative Distribution of Results for a single operator

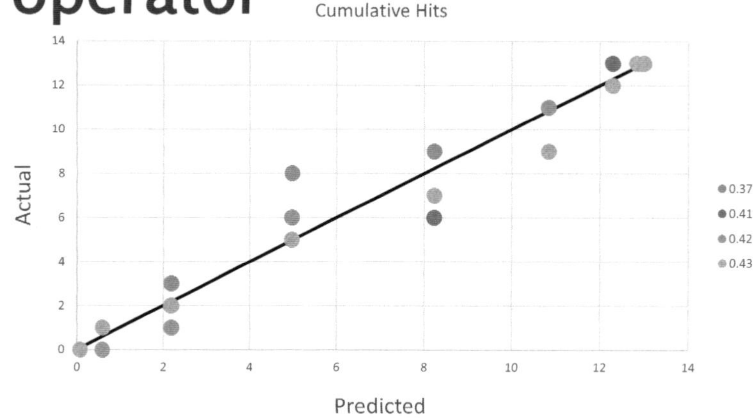

Binomial distribution for p=0.4

Approach

Pulsed Eddy Current

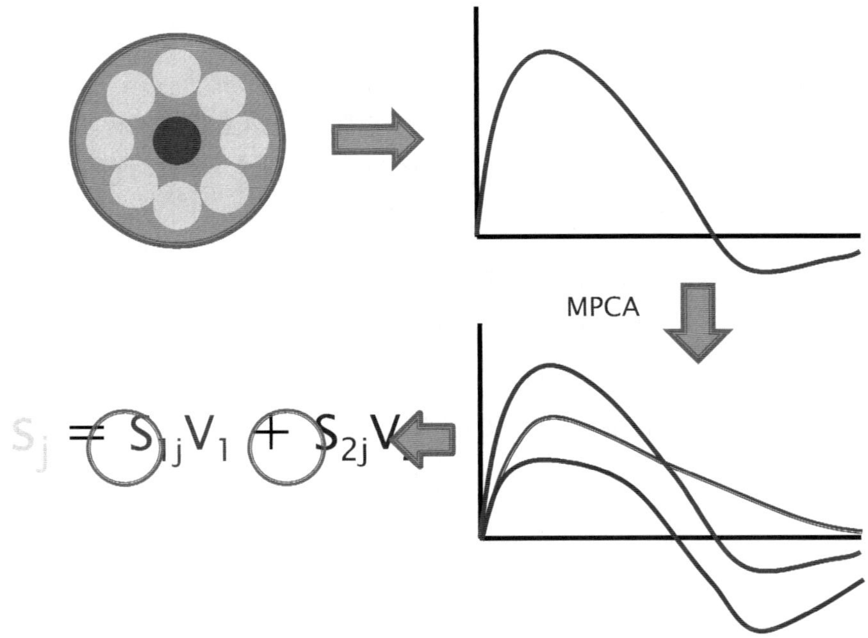

Robust Statistics

Modified Smallest Half Volume

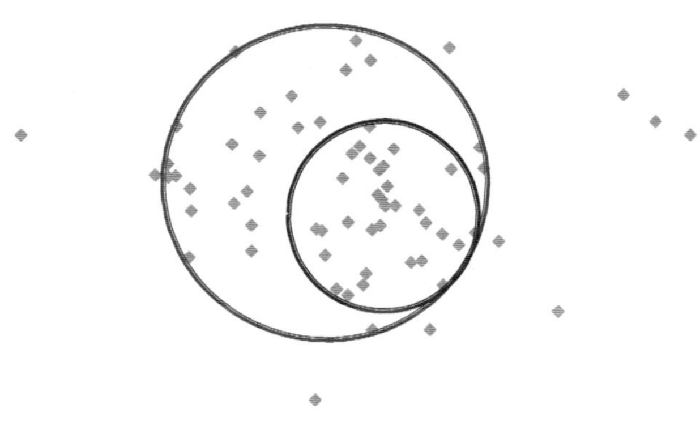

Blind Detection Issues

- How to obtain an uncontaminated covariance matrix and centroid for the blanks?
- Robust Statistics
- Lab samples are defect rich. How to generate large numbers of blanks like the real case?
- Bootstrap Method

Bootstrap

Blanks

S2	S3	S4	S5
0.59337 6	0.04551 9	- 0.09982	- 0.01491
0.50036 4	0.09798 2	-0.0933	- 0.01727
0.44707 1	0.07175 8	- 0.07776	- 0.01358
0.49072 6	0.07744 5	- 0.08257	- 0.01078
0.47631 4	0.05139 5	- 0.07991	-0.0113
0.48483	0.07621 8	- 0.08172	- 0.01324
0.44351 5	0.02831 3	- 0.07137	- 0.01277
0.44633 7	0.05596 7	- 0.07361	- 0.01268
0.49945			

Cracks

S2	S3	S4	S5
- 0.46111	- 0.01135	0.08881 7	- 0.02026
0.30769 7	0.19112 4	0.01930 4	0.00513
0.29056	- 0.05131	- 0.05987	- 0.02391
0.18009 2	-0.0654	- 0.04422	- 0.03478
	-	0.20707	-

Test Data Set

0.00260 6	0.01463 9	- 0.00898	- 0.00704
- 0.08587	0.13199 2	- 0.00031	-0.001
- 0.11465	- 0.02632	0.02628 6	0.00677 2
- 0.33553	- 0.02745	0.00938 1	0.00278 2
-0.0343	- 0.01639	0.01160 7	- 0.00028
0.11737 1	- 0.05911	- 0.01063	- 0.00201
- 0.11154	0.00201	0.02609 4	0.00034 5
0.25988 4	- 0.00134	- 0.01956	- 0.00196
- 0.01007	- 0.02041	-0.0117	- 0.00295
0.08017 6	0.01749 2	0.00521 5	0.01765 6
0.48483	0.07621 8	- 0.08172	- 0.01324
- 0.39145	0.07217 8	0.02476 1	0.00505 7
0.00399 8	- 0.00893	-0.0153	0.00272 3
- 0.03812	0.01479 2	0.01332 2	- 0.00216

AUTHOR INDEX